SpringerBriefs in Energy

SpringerBriefs in Energy presents concise summaries of cutting-edge research and practical applications in all aspects of Energy. Featuring compact volumes of 50 to 125 pages, the series covers a range of content from professional to academic. Typical topics might include:

- A snapshot of a hot or emerging topic
- A contextual literature review
- A timely report of state-of-the art analytical techniques
- An in-depth case study
- A presentation of core concepts that students must understand in order to make independent contributions.

Briefs allow authors to present their ideas and readers to absorb them with minimal time investment.

Briefs will be published as part of Springer's eBook collection, with millions of users worldwide. In addition, Briefs will be available for individual print and electronic purchase. Briefs are characterized by fast, global electronic dissemination, standard publishing contracts, easy-to-use manuscript preparation and formatting guidelines, and expedited production schedules. We aim for publication 8–12 weeks after acceptance.

Both solicited and unsolicited manuscripts are considered for publication in this series. Briefs can also arise from the scale up of a planned chapter. Instead of simply contributing to an edited volume, the author gets an authored book with the space necessary to provide more data, fundamentals and background on the subject, methodology, future outlook, etc.

SpringerBriefs in Energy contains a distinct subseries focusing on Energy Analysis and edited by Charles Hall, State University of New York. Books for this subseries will emphasize quantitative accounting of energy use and availability, including the potential and limitations of new technologies in terms of energy returned on energy invested. The second distinct subseries connected to SpringerBriefs in Energy, entitled Computational Modeling of Energy Systems, is edited by Thomas Nagel, and Haibing Shao, Helmholtz Centre for Environmental Research - UFZ, Leipzig, Germany. This sub-series publishes titles focusing on the role that computer-aided engineering (CAE) plays in advancing various engineering sectors, particularly in the context of transforming energy systems towards renewable sources, decentralized landscapes, and smart grids.

All Springer brief titles should undergo standard single-blind peer-review to ensure high scientific quality by at least two experts in the field.

Alireza Ghadertootoonchi ·
Armaghan Solaimanian · Mehdi Davoudi ·
Moein Moeini Aghtaie

Energy System Modeling and Optimization

A Practical Guide Using Pyomo

Springer

Alireza Ghadertootoonchi
Department of Civil and Mineral Engineering
University of Toronto
Toronto, ON, Canada

Armaghan Solaimanian
Tehran, Iran

Moein Moeini Aghtaie
Tehran, Iran

Mehdi Davoudi
Elmore School of Electrical and Computer Engineering
Purdue University
West Lafayette, IN, USA

ISSN 2191-5520 ISSN 2191-5539 (electronic)
SpringerBriefs in Energy
ISBN 978-3-031-65905-8 ISBN 978-3-031-65906-5 (eBook)
https://doi.org/10.1007/978-3-031-65906-5

This Springer imprint is published by the registered company Springer Nature Switzerland AG
The registered company address is: Gewerbestrasse 11, 6330 Cham, Switzerland

Preface

Nowadays, in the field of energy system studies, it is a well-known stereotype to start by emphasizing that the reason behind the importance of the research is climate change, and this brief book is no exception. Usually, the main solution to tackle the problem of climate change is the integration of renewable energy sources like wind and solar. Nonetheless, another aspect of energy system studies that must not be underestimated is paying attention to people's well-being and satisfying their energy demand reliably. At the same time, they should be encouraged to reduce their energy consumption appropriately. Considering all these factors together for energy system optimization purposes is a complex task and is the subject of the current book.

The book is designed for any reader who is interested in learning more about how to enhance the sustainability and performance of energy systems. No background in energy engineering or energy system studies is needed since the modeling starts from scratch. Furthermore, no background in the Pyomo package is needed, as all the essential functions of Pyomo are first discussed and introduced and then used in the energy system optimization case studies. After reading this book, the reader will be able to model an energy system optimization problem from scratch, acknowledge the complexities of such a problem, describe the potential impacts of different constraints on the performance of the energy system, and demonstrate the advantages and limitations of energy system optimization studies.

The book explores different aspects of energy system optimization, such as the operational cost of power plants, which is directly linked to their fuel consumption, the operational and technical constraints of thermal power plants, the effect of renewable energy and energy storage integration on the energy network, the role of transmission lines, the effect of demand response programs, and the importance of reliability measures when optimizing and scheduling energy systems. To achieve this purpose, the book describes the mathematical foundation of modeling the system in an easy-to-follow but not oversimplified manner. To make the book more accessible to every reader, all the codes are implemented in Python using Pyomo, which are freely available. In addition, to solve the optimization problems, open-source solvers such as GLPK, CBC, and HiGHS are used, which are available online at no cost. Moreover, the codes used for modeling and optimizing the case studies in the book, along with input parameter templates, are available on the authors' GitHub. The reader can download the code, manipulate the parameters, and run multiple cases to see the impact of changes in modeling inputs on the results.

There are a total of five chapters in the book. Chapter 1 is the introduction and provides important information on the energy system supply side and its environmental impacts, as well as other concepts involved in the energy system such as renewables, transmission lines, reliability, and energy self-sufficiency. Chapter 2 is about how to effectively use Pyomo. This chapter has a practical approach to teach the reader how to work with the package and uses a variety of famous optimization problems for this purpose. Chapter 3 covers the mathematical equations that help formulate the optimization problem. The Python code for the energy system optimization problem is presented and discussed in Chap. 4. Finally, Chap. 5 presents the results of different case studies solved using the mentioned Python code.

Toronto, Canada — Alireza Ghadertootoonchi
Tehran, Iran — Armaghan Solaimanian
West Lafayette, USA — Mehdi Davoudi
Tehran, Iran — Moein Moeini Aghtaie

Contents

Chapter 1
Introduction

Abstract Two main parts of an energy system are the supply side and the demand side which are connected by the transmission and distribution lines. Historically, the supply side is equipped with thermal power plants consuming fossil fuel to generate electricity. However, the environmental impact of these power plants is considerable which is one of the main reasons behind the integration of renewable energies in the power system. Considering the demand side, energy consumption is rapidly increasing due to various factors such as population growth, economic development, and climate change. It is important to know the energy system's historical trends as well as its potential future direction. Therefore, this chapter aims to provide necessary background information on the current situation and future predictions of the energy system.

1.1 Introduction

Total final energy consumption of the world has increased steadily during past decades and reached from 194 EJ in 1973 to 418 EJ in 2019, a 115.5% growth over 47 years [1]. According to the International Energy Agency (IEA), the most rapid increase in the share of different energy sources in the overall energy consumption belongs to electricity which shows an approximately 10% growth from 9.5 to 19.7% as depicted in Fig. 1.1. This growth reflects the important role of electricity in final energy consumption.

Electricity consumption has increased due to various reasons, such as the addition of new loads [2], but the main ones are the population growth and the electrification of different devices of end users [3]. Furthermore, due to the important role of electricity in achieving net-zero emissions, which is one of the pivotal sustainable development targets, its share in final energy consumption will continue growing in the future. Based on the 2050 net-zero emission scenario, the share of electricity in the final energy demand should increase by 4% annually [4]. In addition, the electricity supply system is transitioning to a more distributed system from the conventional vertical structure in which several large-capacity and centralized power plants exist [5]. The

A. Ghadertootoonchi et al., *Energy System Modeling and Optimization*,
SpringerBriefs in Energy, https://doi.org/10.1007/978-3-031-65906-5_1

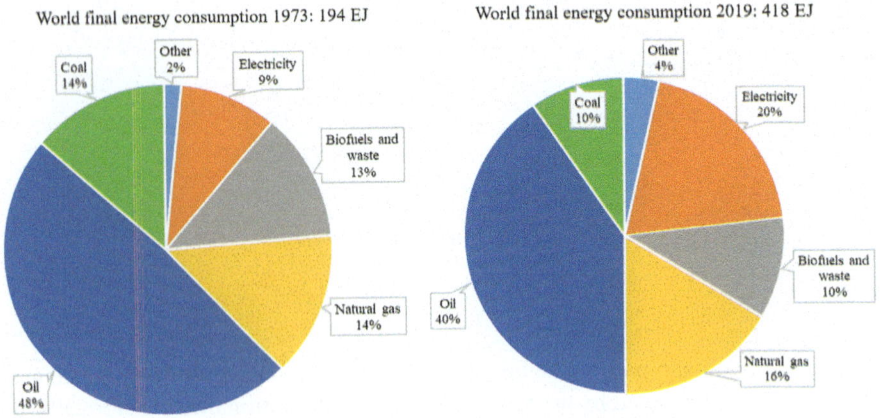

Fig. 1.1 Share of different energy sources in global energy consumption [1]

main motivations to transform the electricity generation and transmission system to a more distributed one are (i) deregulation of the system and introduction of the electricity markets which increases the efficiency of the electricity system and reduces the energy subsidies [6–8], (ii) growing trend of renewable energies integration to increase the sustainability of the system [9, 10], (iii) resiliency improvement of the electricity supply system [11, 12].

The deregulation of the electricity system can lead to cost reduction and more efficient resource allocation [13–15]. In addition, a deregulated and competitive electricity market can assist the energy transition goals such as effective implementation of carbon pricing and renewable energy integration [16, 17]. Renewable energy sources play a crucial role in the future electricity supply system and their share in the energy system is growing as shown in Fig. 1.2 [18]. Based on the IEA, total renewable electricity generation in 2022 experienced a 600 TWh growth, compared to 2021, and reached 8500 TWh with the majority coming from wind and solar technologies [19]. Therefore, when developing a large-scale model for the energy system, it is important not to neglect the impact of renewable energy integration.

The third reason behind the transition of the electricity system to a more distributed version is increasing the system's resiliency. Electricity systems can face several challenges such as natural disasters [20] and extreme weather events such as heat waves [21]. The central or traditional electricity system is more vulnerable to these conditions and can fail to provide electricity reliably [22]. However, decentralized systems that can operate independently from the central plants are emerging alternatives to address this issue [23]. Interested readers can find the resiliency indicators and their definitions in [21].

Considering the above, it can be seen that the structure of the power system is changing, and at the same time, electricity consumption is growing. Such a condition raises the need for increasing the accuracy and complexity of power system modeling. This can be done through the integration of new constraints into the simulation and

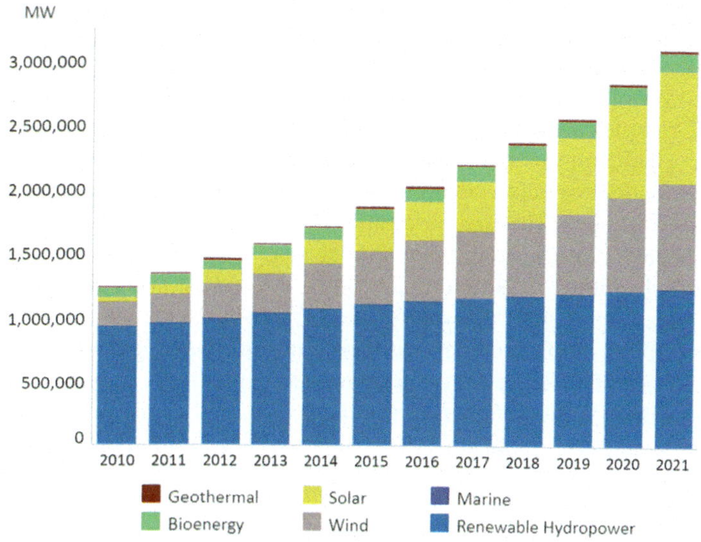

Fig. 1.2 The growing trend of renewable energy capacity from 2010 to 2021 [18]

optimization models. The following sections categorize these constraints into supply side, demand side, and transmission and distribution and provide a background for each of them.

1.2 Supply Side

In the new structure of the electricity system, there are various electricity generation units with different capacities and characteristics [24]. Some of these units are dispatchable, like traditional thermal power plants, and some are non-dispatchable, like wind and solar power plants [25, 26]. Dispatchable units can change their generation capacity from one period to another and respond to demand changes. A good example of dispatchable units is gas turbines with a high ramp up and ramp down rate and adjustable capacity. On the other hand, such characteristics do not exist for renewable energies such as solar, and reducing the output of the solar power plants is done through curtailment. Modeling and optimization of thermal power plants (dispatchable) and renewable sources (non-dispatchable) have been the subjects of multiple studies which are mentioned in the following sub-sections.

1.2.1 Thermal Power Plants

Thermal power plants are dispatchable, provide inertia, and maintain energy security [27]. Examples are gas or steam turbines (or both if the plant is a combined cycle). The studies in this realm have tried to model these power plants by developing mathematical models as realistically as possible considering a wide variety of operational constraints. Thermal power plants have a variety of technical constraints that must be taken into account when modeling power systems. The aim is to find the optimal scheduling of thermal units so that the operational costs are minimized (Economic dispatch) [28, 29]. Usually, the dispatching problem is formulated as a non-convex optimization problem and then reformulated as a mixed-integer linear program (MILP) [30]. However, other optimization methods, such as fuzzy optimal search methods [31], evolutionary algorithms [32], and swarm intelligence-based methods [33], are also applicable. Nevertheless, these units burn fossil fuels and emit GHG; More than 60% of global electricity generation depends on burning coal, oil, and natural gas in thermal power plants [34]. Therefore, special attention needs to be devoted to the environmental impacts of these fossil fuel-based units. An overview of the existing techniques for doing so is provided in the next subsection.

1.2.2 Environmental Impacts

Due to the significant contribution of the electricity supply system to climate change, one should consider various objectives alongside the energy costs in the optimization problem associated with energy sources. Energy-related GHG emissions in 2022 were 36.8 Gt of which 14.65 Gt is directly related to the power sector and more than one-third is the result of burning coal as an energy source. Figure 1.3 shows global energy-related GHG emissions divided by energy sources [35].

Such a strong contribution of the power sector in GHG emissions implies that the power plant owners should minimize not only the costs but also the environmental impacts. Among the methods available for assessing the impact of carbon emissions in optimization models, the consideration of emission-related monetary costs is more prevalent.

One of the options among many to reflect the monetary cost of emissions is through carbon taxing which assigns a monetary value to pollution. Another is through an emission trading system (ETS). ETS provides a market in which low emitters of carbon can sell their emission allowance to those with higher emissions. The market price for every unit of emission determines the carbon price. Regardless of the mechanism, one should note that every unit of GHG emissions has a cost that has to be considered when optimizing the energy system. According to the World Bank carbon pricing dashboard, currently, 73 carbon pricing initiatives are implemented around the world covering 11.7 Gt of CO_2 equivalent per year which is equal to 23% of the overall GHG emissions. Figure 1.4 shows the global map of carbon pricing initiatives

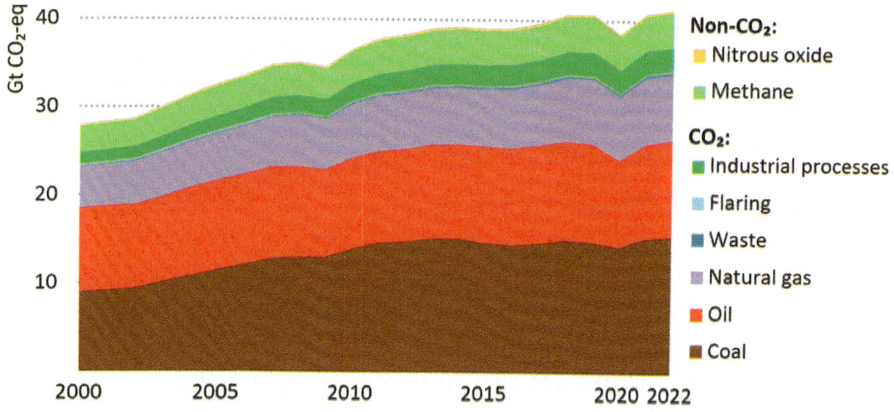

Fig. 1.3 Global energy-related GHG emissions divided by energy sources [35]

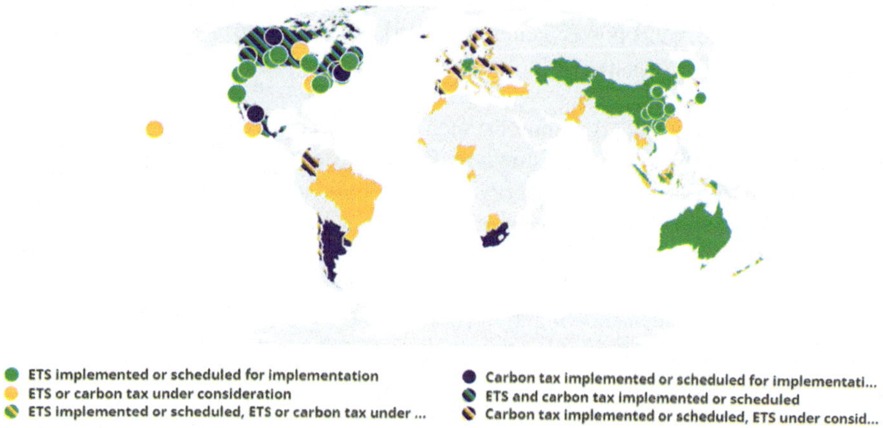

Fig. 1.4 Global carbon pricing initiatives [36]

as of March 2023 in which the price could be as low as 0.82 USD per tCO_2e for Ukraine or as high as 155.87 USD per tCO_2e for Washington CCA[1] [36].

Such a condition opens a new category of power systems optimization problems called eco-environmental optimization in which the aim is to minimize the economic costs as well as emission costs. To this end, two methods are available to analyze the operational and environmental costs [28]. The first approach imposes an upper bound on the emissions, and its objective is to minimize the cost. It is difficult to analyze the cost-emission trade-off in this method [37]. In the second approach, a multi-objective-MILP optimization model is used to consider both economic and GHG emissions costs [38–40].

---

[1] Climate Commitment Act.

Considering emission costs provides an incentive for the decision-makers to reduce their emissions. Nonetheless, a permanent solution is the integration of renewables into the power system. The subsequent section delves into this matter.

1.2.3 Renewable Power Plants

Studies used different mathematical models for renewable energy technologies such as wind and solar to effectively integrate them into the electricity system. PV panels can be modeled by considering a fixed conversion efficiency [41, 42] or the impact of ambient temperature and solar radiation intensity on their efficiency. The higher the ambient temperature, the lower the conversion efficiency and panel output [43, 44]. Beta probability density function (PDF) can be used to incorporate the uncertainty of solar power generation [45]. Power output of wind turbines is modeled using a piecewise function of wind velocity [41, 42] and Weibull PDF is used to model the uncertainty of wind power generation [45]. Furthermore, PV and wind power generation can be extracted from external resources and used in the optimization model [46, 47].

One problem with energy sources such as wind and solar is that they depend on weather conditions and are known as variable renewable energies (VRE). This brings up the need for electric energy storage (EES) units. The impact of EES on the optimal operation of energy systems has been the subject of multiple studies [48]. It is shown that EES integration increases PV utilization by up to 60% as the system stores excess energy during the day and releases it during times of need [49]. Furthermore, it has been shown that the existence of EES can increase the economic and environmental benefits [50]. EES systems can also reduce the impact of power fluctuation and improve grid reliability [51]. The next subsection explores the crucial aspect of reliability on the power system's supply side.

1.2.4 Reliability

The power system's supply side has to have the ability to deal with various extreme conditions under which some of the components may fail; Yet, the electricity network must be able to satisfy the demand. For instance, if one power plant fails, the other power plants must be able to cover the lost capacity as quickly as possible. Such a concept is called $N - 1$ contingency and has been used in previous studies [52]. $N - 1$ contingency analysis is a useful technique for assessing the reliability of power systems and can aid in developing plans for enhancing system functionality and lowering the chance of blackouts [53]. For instance, authors in [54] used the concept of $N - 1$ contingency and conditional probabilities of system failure to analyze the reliability of the system. An example of economic dispatch with contingency constraints can be found in [55].

Overall, while it is crucial to model and optimize both dispatchable and non-dispatchable plants considering the above complexities, it is important to note that the supply side alone does not encompass the entirety of the electricity system. The system comprises several other vital sections like transmission lines and the demand side which are the subject of the next sections.

1.3 Transmission and Distribution System

Transmission and distribution system transfers electricity from the supply to the demand and play a crucial role in the electricity system [56, 57]. Therefore, developing an optimization model solely based on the supply side does not reflect system interdependencies.

To address these interdependencies, some studies developed economic dispatch models considering the impact of transmission and distribution networks. For instance, authors in [58] used direct search methods to solve economic dispatch problems for one and two-area power systems considering transmission line capacity. The results show that transmission line capacity affects the operational cost. Similarly, authors in [59] considered transmission line capacity as one of the constraints of the system and added transmission cost to the economic dispatch problem objective function. The importance of considering transmission and distribution lines in the optimization of energy systems in the presence of distributed energy resources is discussed in [60]. In addition to transmission lines, the demand side is another important part of the electricity system which will be discussed in the following.

1.4 Demand Side

The demand side is another important part of the electricity system. In many cases, it is considered that the power plants, either renewable or fossil fuel-based, have to satisfy a static demand. However, in deregulated and distributed electricity systems more and more consumers join the demand response programs (DR) and change their behavior according to the market condition which is directly affected by the operation of power plants. DR programs can address problems such as peak demand issues, high generation costs, emissions, and grid congestion [61]. On top of that, electricity market deregulation opens the road for interested consumers to produce electricity and sell it in the market resulting in the formation of a new category of end users called prosumers [62]. As a result, the assumption of static demand over a period of time does not reflect the true behavior of the demand side any more and more sophisticated modeling and optimization methods are needed to encode this new behavior into the mathematical models.

Numerous studies have been conducted to assess the influence of DR on the optimal functioning of energy systems. Notably, it has been found that DR can lead to cost savings, particularly by avoiding network-delivered electricity costs and reducing capital investment [63]. Moreover, the reliability of power systems during generation failure can be ameliorated using DR programs [64]. They can also facilitate renewable energy integration and alleviate fossil fuel dependency [65]. In addition, DR programs can also ameliorate regional energy self-sufficiency which will be discussed in the next section.

1.5 Energy Self-Sufficiency

One of the characteristics of distributed electricity networks which is related to supply, demand, and transmission systems is energy self-sufficiency; Meaning that every region in the distributed network should preferably have the ability to satisfy a part of or all of its energy demand over a specific period without being dependent on other regions [66]. This constraint ensures energy security within each region and limits the effect of energy system failure in a region to the same region. Studies that considered energy self-sufficiency or net zero energy constraint can be divided into two main groups, small-scale systems (e.g., for a single residential, industrial, or commercial building) [67] and large-scale systems (e.g., net zero energy neighborhoods and districts) [68].

Despite the importance of net-zero energy buildings, these studies mainly focus on the performance of the building, not its effects on the electricity grid [69]. To bridge the mentioned gap, net-zero energy neighborhoods (NZEN), districts (NZED), or even regions (NZER) must be analyzed. These studies impose the net-zero energy constraint on the neighborhood or district levels since the result of the building-level modeling could be sub-optimal from the large-scale (e.g., grid) perspective [70].

NZENs can utilize different energy conversion technologies. For instance, authors in [71] proposed an energy system consisting of PV, wind turbine, solar thermal collector, waste to energy (W2E), and seasonal thermal storage to design an NZEN in Calgary, Canada. Another research investigated the role of solar energy in providing electricity and the heat demand of a neighborhood [72]. The applicability of solar energy in NZEDs of the Mediterranean region is studied in [73].

1.6 Concluding Remarks

The mentioned factors increase the complexity of the energy system and add several constraints and limitations to the operation of the system. As a result, modeling and optimization of electricity systems becomes more difficult. A majority of optimization problems in this area are formulated as MILP models; Therefore, this book aims to provide a step-by-step guide for developing a MILP optimization model for a distributed electricity system that addresses the above complexities. A variety of different existing and emerging technologies and trends such as thermal power plants, variable renewable energies, energy storage technologies, transmission/distribution lines, demand side management, environmental factors, reliability, and energy self-sufficiency are considered in this book and the final model will cover all these aspects.

This is accomplished through the use of a dynamic, multi-region eco-environmental optimization model that has been developed using a graph. In this graph, electricity supply and demand locations are represented as nodes (i.e., vertices), and the transmission lines as edges. This framework enables the integration of existing data into the mathematical model, facilitating the optimization of the system. The problem is then modeled using Pyomo which is an open-source advanced mathematical programming language (AMPL) and solved using different open-source solvers such as GLPK, CBC, and HiGHS which is a state-of-the-art open-source solver.

The rest of the book is organized as follows. Chap. 2 introduces Pyomo and guides the reader on how to use it. After reading Chap. 2, the reader can develop a Python code for an optimization problem and will have the necessary knowledge to solve energy system optimization problems.

Chapter 3 is devoted to energy system modeling. The chapter describes the mathematical formulation of different technologies used in energy systems such as thermal power plants and their technical characteristics, electricity storage systems, and renewable energies such as wind and solar, hydroelectric units, and combined heating and power generation units. In addition, the mathematical formulation of reliability, load loss, and demand-side flexibility is discussed. Finally, some of the linearization techniques used in this domain of study along with the impact of variable renewable energy uncertainty on problem formulation. After reading this chapter, the reader can develop the mathematical model of an energy system and solve it using Pyomo. In addition, the reader will be able to interpret and analyze the results.

Finally, Chap. 4 is devoted to a case study in which most of the technologies introduced in the previous chapter are involved. It begins with a simple optimization problem; The complexity of the energy system will increase in an easy-to-follow approach so that the reader can vividly see how to use the models and constraints introduced in the previous chapter to optimize a complex energy system. In addition, a set of valid data will be used in the case study to make it as realistic as possible.

References

1. International Energy Agency (2023) Key world energy statistics 2021. IEA. https://www.iea.org/reports/key-world-energy-statistics-2021/final-consumption. Accessed 07 Nov 2023
2. Akbarnavasi H, Ledari MB, Ghadertootoonchi A (2023) An optimization strategy for enhancing energy consumption performance in digital currency miner's building. Energy Convers Econ 4(2):137–157. https://doi.org/10.1049/enc2.12085
3. Ghadertootoonchi A, Davoudi M, Moeini-Aghtaie M, Rahmani-Andebili M (2023) Role of EVs in the optimal operation of multicarrier energy systems BT: planning and operation of electric vehicles in smart grids. In: Rahmani-Andebili M (ed) Springer, Cham, pp 69–117

4. International Energy Agency (2023) Tracking clean energy progress 2023. IEA. https://www.iea.org/reports/tracking-clean-energy-progress-2023. Accessed 07 Nov 2023
5. Chwieduk D, Bujalski W, Chwieduk B (2020) Possibilities of transition from centralized energy systems to distributed energy sources in large polish cities. Energies 13(22):6007. https://doi.org/10.3390/en13226007
6. Bojnec Š, Križaj A (2021) "Electricity markets during the liberalization: the case of a European Union country. Energies 14(14):4317. https://doi.org/10.3390/en14144317
7. Sorknæs P et al (2020) Smart energy markets: future electricity, gas and heating markets. Renew Sustain Energy Rev 119:109655. https://doi.org/10.1016/j.rser.2019.109655
8. Ghadertootoonchi A, Bararzadeh M, Fani M (2020) The effect of energy subsidies on the sustainability of economy, society and environment: a case study of Iran. Econ Policy Energy Environ 2:3280. https://doi.org/10.3280/EFE2020-002005
9. Sinsel SR, Riemke RL, Hoffmann VH (2020) Challenges and solution technologies for the integration of variable renewable energy sources—a review. Renew Energy 145:2271–2285. https://doi.org/10.1016/j.renene.2019.06.147
10. Johannsen RM, Mathiesen BV, Kermeli K, Crijns-Graus W, Østergaard PA (2023) Exploring pathways to 100% renewable energy in European industry. Energy 268:126687. https://doi.org/10.1016/j.energy.2023.126687
11. Mahzarnia M, Moghaddam MP, Baboli PT, Siano P (2020) A review of the measures to enhance power systems resilience. IEEE Syst J 14(3):4059–4070. https://doi.org/10.1109/JSYST.2020.2965993
12. Davoudi M, Barmayoon MH, Moeini-Aghtaie M (2023) Multi-objective optimal planning of a residential energy hub based on multi-objective particle swarm optimization algorithm. IET Gener Transm Distrib 17(10):2435–2448. https://doi.org/10.1049/gtd2.12820
13. MacKay A, Mercadal I (2021) Deregulation, market power, and prices: evidence from the electricity sector. https://doi.org/10.2139/ssrn.3793305
14. Hughes JE, Lange I (2020) Who (else) benefits from electricity deregulation? Coal prices, natural gas, and price discrimination. Econ Inq 58(3):1053–1075. https://doi.org/10.1111/ecin.12896
15. Chawla M, Pollitt MG (2013) Global trends in electricity transmission system operation: where does the future lie? Electr J 26(5):65–71. https://doi.org/10.1016/j.tej.2013.05.004
16. Davoudi M, Moeini-Aghtaie M (2022) Local energy markets design for integrated distribution energy systems based on the concept of transactive peer-to-peer market. IET Gener Transm Distrib 16(1):41–56. https://doi.org/10.1049/gtd2.12274
17. Wu L (2023) Handbook of energy law in the low-carbon transition. In: Bellantuono G, Godden L, Mostert H, Wiseman H, Zhang H (eds) Handbook of energy law in the low-carbon transition. De Gruyter, pp 123–144
18. International Renewable Energy Agency (2023) Historical capacity of main renewable elctricity technologies. IRENA. https://www.irena.org/Data/View-data-by-topic/Capacity-and-Generation/Statistics-Time-Series. Accessed 07 Nov 2023
19. Electricity Market Report 2023. https://doi.org/10.1787/f0aed4e6-en
20. Driesen J, Katiraei F (2008) Design for distributed energy resources. IEEE Power Energy Mag 6(3):30–40. https://doi.org/10.1109/MPE.2008.918703
21. Liu G, Jiang T, Ollis TB, Li X, Li F, Tomsovic K (2020) Resilient distribution system leveraging distributed generation and microgrids: a review. IET Energy Syst Integr 2(4):289–304. https://doi.org/10.1049/iet-esi.2019.0134
22. Ratnam EL, Baldwin KGH, Mancarella P, Howden M, Seebeck L (2020) Electricity system resilience in a world of increased climate change and cybersecurity risk. Electr J 33(9):106833. https://doi.org/10.1016/j.tej.2020.106833
23. Bouffard F, Kirschen DS (2008) Centralised and distributed electricity systems. Energy Policy 36(12):4504–4508. https://doi.org/10.1016/j.enpol.2008.09.060
24. Davoudi M, Moeini-Aghtaie M, Ghorani R (2021) Developing a new framework for transactive peer-to-peer thermal energy market. IET Gener Transm Distrib 15:12150. https://doi.org/10.1049/gtd2.12150

25. Baroni M (2022) The integration of non-dispatchable renewables BT. In: Hafner M, Luciani G (eds) The Palgrave handbook of international energy economics. Springer, Cham, pp 269–299
26. Joskow PL (2011) Comparing the costs of intermittent and dispatchable electricity generating technologies. Am Econ Rev 101(3):238–241. https://doi.org/10.1257/aer.101.3.238
27. International Energy Agency (2021) The role of low-carbon fuels in the clean energy transitions of the power sector. https://doi.org/10.1787/a92fe011-en
28. Abido MA (2009) Multiobjective particle swarm optimization for environmental/economic dispatch problem. Electr Power Syst Res 79(7):1105–1113. https://doi.org/10.1016/j.epsr.2009.02.005
29. Zivic Djurovic M, Milacic A, Krsulja M (2012) A simplified model of quadratic cost function for thermal generators. In: Proceedings of the 23rd DAAAM international symposium on intelligent manufacturing automics, pp 25–28. https://doi.org/10.2507/23rd.daaam.proceedings.006
30. Shahidehpour M, Yamin H, Li Z (2002) PriceBased unit commitment. In: Market operations in electric power systems: forecasting, scheduling, and risk management. IEEE, pp 115–160
31. Srinivasan D, Chang CS, Liew AC (1994) Multiobjective generation scheduling using fuzzy optimal search technique
32. Abido MA (2003) A novel multiobjective evolutionary algorithm for environmental/economic power dispatch. Electr Power Syst Res 65(1):71–81. https://doi.org/10.1016/S0378-7796(02)00221-3
33. Gupta GK, Goyal S (2017) Particle swarm intelligence based dynamic economic dispatch with daily load patterns including valve point effect. In: Proceedings of the 2017 3rd international conference on condition assessment techniques in electrical systems (CATCON), pp 83–87. https://doi.org/10.1109/CATCON.2017.8280189
34. International Energy Agency (IEA) (2022) Energy statistics data browser. IEA. https://www.iea.org/data-and-statistics/data-tools/energy-statistics-data-browser?country=WORLD&fuel=Energysupply&indicator=ElecGenByFuel. Accessed 29 Dec 2023
35. International Energy Agency (2023) CO_2 emissions in 2022. https://doi.org/10.1787/12ad1e1a-en
36. The World Bank (2023) Carbon pricing dashboard. The World Bank Group. https://carbonpricingdashboard.worldbank.org/map_data
37. Granelli GP, Montagna M, Pasini GL, Marannino P (1992) Emission constrained dynamic dispatch. Electr Power Syst Res 24(1):55–64. https://doi.org/10.1016/0378-7796(92)90045-3
38. Zhang R, Zhou J, Mo L, Ouyang S, Liao X (2013) Economic environmental dispatch using an enhanced multi-objective cultural algorithm. Electr Power Syst Res 99:18–29. https://doi.org/10.1016/j.epsr.2013.01.010
39. Talbi EH, Abaali L, Skouri R, El Moudden M (2020) Solution of economic and environmental power dispatch problem of an electrical power system using BFGS-AL algorithm. Proced Comput Sci 170(2019):857–862. https://doi.org/10.1016/j.procs.2020.03.144
40. Davoudi M, Jooshaki M, Moeini-Aghtaie M, Hossein-Barmayoon M, Aien M (2022) Developing a multi-objective multi-layer model for optimal design of residential complex energy systems. Int J Electr Power Energy Syst 138:107889. https://doi.org/10.1016/j.ijepes.2021.107889
41. Maleki A, Askarzadeh A (2014) Optimal sizing of a PV/wind/diesel system with battery storage for electrification to an off-grid remote region: a case study of Rafsanjan, Iran. Sustain Energy Technol Assess 7:147–153. https://doi.org/10.1016/j.seta.2014.04.005
42. Sharafi M, ELMekkawy TY (2014) Multi-objective optimal design of hybrid renewable energy systems using PSO-simulation based approach. Renew Energy 68:67–79. https://doi.org/10.1016/j.renene.2014.01.011
43. Huang Z, Xie Z, Zhang C, Chan SH, Milewski J (2018) PV-hydrogen-retired EV battery hybrid energy system 1 Introduction, p 31
44. Luo Y, Zhang X, Yang D, Sun Q (2020) Emission trading based optimal scheduling strategy of energy hub with energy storage and integrated electric vehicles. J Mod Power Syst Clean Energy 8(2):267–275

45. Mokaramian E, Shayeghi H, Sedaghati F, Safari A, Alhelou HH (2022) An optimal energy hub management integrated EVs and RES based on three-stage model considering various uncertainties. IEEE Access 10:17349–17365. https://doi.org/10.1109/ACCESS.2022.3146447
46. Ghadertootoonchi A, Moeini-Aghtaie M, Davoudi M (2023) A hybrid linear programming-reinforcement learning method for optimal energy hub management. IEEE Trans Smart Grid 14(1):157–166. https://doi.org/10.1109/TSG.2022.3197458
47. Solaimanian A, Roshandel R (2023) Effect of PV roof coverage on the lighting availability, heating, and cooling demands for a Venlo greenhouse in Tehran. In: Proceedings of the 2023 8th international conference on technology and energy management (ICTEM), pp 1–5. https://doi.org/10.1109/ICTEM56862.2023.10083896
48. Shaw-Williams D, Susilawati C, Walker G, Varendorff J (2020) Towards net-zero energy neighbourhoods utilising high rates of residential photovoltaics with battery storage: a techno-economic analysis. Int J Sustain Energy 39(2):190–206. https://doi.org/10.1080/14786451.2019.1668394
49. Mavrigiannaki A et al (2021) Zero energy concept at neighborhood level: a case study analysis. Sol Energy Adv 1:100002. https://doi.org/10.1016/j.seja.2021.100002
50. Sameti M, Haghighat F (2018) Integration of distributed energy storage into net-zero energy district systems: optimum design and operation. Energy 153:575–591. https://doi.org/10.1016/j.energy.2018.04.064
51. Scholczova P (2023) Energy storage efficiency. In: Xu PJ, Yang PB (eds) IntechOpen, Rijeka
52. Faraji J, Hashemi-Dezaki H, Ketabi A (2021) Stochastic operation and scheduling of energy hub considering renewable energy sources' uncertainty and N-1 contingency. Sustain Cit Soc 65:102578. https://doi.org/10.1016/j.scs.2020.102578
53. Prabhakar PSV, Krishan R, Pullaguram DR (2022) Static security assessment of large power systems under N–1–1 contingency. In: Proceedings of the 2022 22nd national power systems conference (NPSC), pp 35–40. https://doi.org/10.1109/NPSC57038.2022.10069705
54. Alshammari B (2019) Evaluation of power system reliability levels for (n-1) outage contingency. Int J Adv Appl Sci 6:68–74
55. Aganagic M, Awobamise B, Raina G, McCartney AI (1997) Economic dispatch with generation contingency constraints. IEEE Trans Power Syst 12(3):1229–1236. https://doi.org/10.1109/59.630465
56. Albadi MH, Al-Farsi FN, Hosseinzadeh N, Al-Badi A (2018) Effect of considering transmission losses in economic dispatch: a case study of Oman's main interconnected system. J Eng Res 15(1):1–13
57. Moreira FS, Ohishi T, Filho JIS (2006) Influence of the thermal limits of transmission lines in the economic dispatch. In: Proceedings of the 2006 IEEE power engineering society general meeting, p 6. https://doi.org/10.1109/PES.2006.1709590
58. Zarei M, Roozegar A, Kazemzadeh R, Kauffmann JM (2007) Two area power systems economic dispatch problem solving considering transmission capacity constraints. Int J Energy Power Eng 1:1267–1272
59. Nwulu N (2017) Dynamic economic dispatch of electric power incorporating transmission line costs. In: Proceedings of the 2017 international conference on energy, communication, data analytics and soft computing (ICECDS), pp 217–221. https://doi.org/10.1109/ICECDS.2017.8389817
60. Zhou X, Chang C-Y, Bernstein A, Zhao C, Chen L (2021) Economic dispatch with distributed energy resources: co-optimization of transmission and distribution systems. IEEE Control Syst Lett 5(6):1994–1999. https://doi.org/10.1109/LCSYS.2020.3044542
61. Jordehi AR (2019) Optimisation of demand response in electric power systems, a review. Renew Sustain Energy Rev 103:308–319. https://doi.org/10.1016/j.rser.2018.12.054
62. Parag Y, Sovacool BK (2016) Electricity market design for the prosumer era. Nat Energy 1(4):32. https://doi.org/10.1038/NENERGY.2016.32
63. Ihsan A, Jeppesen M, Brear MJ (2019) Impact of demand response on the optimal, techno-economic performance of a hybrid, renewable energy power plant. Appl Energy 238:972–984. https://doi.org/10.1016/j.apenergy.2019.01.090

64. Aghaei J, Alizadeh M-I, Siano P, Heidari A (2016) Contribution of emergency demand response programs in power system reliability. Energy 103:688–696. https://doi.org/10.1016/j.energy.2016.03.031
65. Tahir MF, Haoyong C, Idris II, Larik NA, Adnan SU (2018) Demand response programs significance, challenges and worldwide scope in maintaining power system stability. Int J Adv Comput Sci Appl 9(6):124–131
66. Engelken M, Römer B, Drescher M, Welpe I (2016) Transforming the energy system: why municipalities strive for energy self-sufficiency. Energy Policy 98:365–377. https://doi.org/10.1016/j.enpol.2016.07.049
67. Omrany H, Chang R, Soebarto V, Zhang Y, Ghaffarianhoseini A, Zuo J (2022) A bibliometric review of net zero energy building research 1995–2022. Energy Build 262:111996. https://doi.org/10.1016/j.enbuild.2022.111996
68. Ullah KR, Prodanovic V, Pignatta G, Deletic A, Santamouris M (2021) Technological advancements towards the net-zero energy communities: a review on 23 case studies around the globe. Sol Energy 224:1107–1126. https://doi.org/10.1016/j.solener.2021.06.056
69. Amaral AR, Rodrigues E, Rodrigues-Gaspar A, Gomes Á (2018) Review on performance aspects of nearly zero-energy districts. Sustain Cit Soc 43:406–420. https://doi.org/10.1016/j.scs.2018.08.039
70. Kazmi H, Keijsers M, Mehmood F, Miller C (2022) Energy balances, thermal performance, and heat stress: disentangling occupant behaviour and weather influences in a Dutch net-zero energy neighborhood. Energy Build 263:112020. https://doi.org/10.1016/j.enbuild.2022.112020
71. Hachem-Vermette C, Singh K (2021) Analysis of urban energy resources to achieve net-zero energy neighborhoods. Front Sustain Cit 3:1–10. https://doi.org/10.3389/frsc.2021.663256
72. Hachem-Vermette C (2022) Role of solar energy in achieving net zero energy neighborhoods. Sol Energy Adv 2:100018. https://doi.org/10.1016/j.seja.2022.100018
73. Guarino F, Tumminia G, Longo S, Mistretta M, Bilotta R, Cellura M (2016) Energy planning methodology of net zero energy solar neighborhoods in the Mediterranean basin. Sci Technol Built Environ 22:5656. https://doi.org/10.1080/23744731.2016.1195656

Chapter 2
How to Use Pyomo

Abstract This chapter provides essential knowledge about how to use the Pyomo optimization library in Python programming language; Basic knowledge of Python is required to follow the codes but the reader is not required to be familiar with the Pyomo library. It starts with an explanation of the main parts of an optimization problem which are variables, bounds, sets, objective function(s), and constraints. After that, six sample optimization problems that are categorized into four groups are presented and solved using Pyomo. All the required concepts to use this book are covered in this chapter.

2.1 Introduction

The focus of this book is on linear optimization problems, and from now on, the optimization problem refers to a linear optimization model. In simple words, linear optimization is a problem in which the relationship between the decision variables (DVs) is completely linear. For instance, the DVs cannot be multiplied together, be exponentiated, and be the input of a nonlinear function such as sin. The expected outcome of this chapter is the ability to develop the mathematical model of a problem, identify different parts of the optimization model, and formulate these parts using the Pyomo library in Python. Different parts of the optimization model are defined in the following section. For a more comprehensive Pyomo tutorial, consider reading the official reference book [1].

2.2 Structure of an Optimization Problem

An optimization problem, whether linear or nonlinear, consists of three main parts; First is the decision variables which are the variables that are optimized through the optimization problem. These variables can be bounded (i.e., have upper and/or lower bounds) or unbounded. Second is the objective function which is a function of

A. Ghadertootoonchi et al., *Energy System Modeling and Optimization*, SpringerBriefs in Energy, https://doi.org/10.1007/978-3-031-65906-5_2

decision variables and is maximized or minimized with regard to the optimal values of the decision variables. Third is the constraints; These are the limitations and rules that should be satisfied considering different values for the decision variables. The overall structure of an optimization problem is brought in Eqs. (2.1)–(2.2).

$$Objective : \min f(X) \tag{2.1}$$

$$AX \leq B \tag{2.2}$$

$$A'X = B' \tag{2.3}$$

$$X_{lb} \leq X \leq X_{ub} \tag{2.4}$$

where X is the vector of decision variables, $f(X)$ is the objective function, A is the multipliers of inequality constraints, B is the right-hand side of the inequality constraints, A' is the multipliers of the equality constraints, B' is the right-hand side of the equality constraints, and X_{lb} and X_{ub} are the decision variables lower and upper bounds respectively. All the elements of A, B, A', B', X_{lb}, and X_{ub} are constant parameters of the problem. The following three subsections explain the mentioned parts in more detail. Before that, it is important to know the possible outcomes of solving an optimization problem.

The main possible results of solving an optimization problem are infeasible, unbounded, suboptimal, and optimal. Infeasibility happens when the constraints are inconsistent; Meaning that one constraint cannot be satisfied as long as the other exists. For instance, if x is a decision variable, having two constraints of $x > 10$ and $x < 5$ leads to infeasibility since x cannot be greater than 10 and less than 5 at the same time. The optimization problem is unbounded when the objective function can be minimized or maximized to reach $-\infty$ or $+\infty$. This happens when one or multiple decision variable(s) have no lower or upper bounds. Finally, if the problem has a set of values for the decision variables that satisfy the constraint, and do not return $-\infty$ or $+\infty$ for the objective function, the solution is feasible and can be either suboptimal or optimal. Optimality happens in some cases such as a convex problem. Suboptimality also happens in some cases such as when the solving procedure stops after reaching a certain number of iterations or time limitation and reports the results; In such a case, the reported results might not be optimal but be suboptimal. The coming section demonstrates how to define variables and use them in Pyomo.

2.2.1 *Variables and Sets*

After installing Pyomo by following the installation guide [2], one can define a decision variable using the code presented in Box 2.1.

Box 2.1

```
1 import pyomo.environ as pyo
2 model = pyo.ConcreteModel()
3 model.x = pyo.Var()
```

Line 1 imports the Pyomo environment; Line 2 defines a model in which all the variables, constraints, and the objective function will be stored. Finally, line 3 defines a decision variable called x. By default, x accepts real values and has no upper or lower bounds. To add bounds and specify the domain, line 3 should be changed to the code shown in Box 2.2.

Box 2.2

```
1 model.x = pyo.Var(within = pyo.NonNegativeReals, bounds =
(0,10))
2 model.y = pyo.Var(within = pyo.PositiveReals, bounds =
(0,None), initialize = 3)
```

Line 1 in Box 2.2 indicates that the domain of x, specified by `within` keyword, is nonnegative real numbers or $[0, +\infty)$. The lower bound of x is 0 and the upper bound is 10. Moreover, a new variable, y, is defined which is positive or $(0, +\infty)$. Its lower bound is 0 and it has no upper bound (Note that **None** is a Python keyword). In addition, one can assign an initial value to the variable using the `initialize` keyword as shown in line 2 of Box 2.2. Possible values that can be passed to `within` are brought in Table 2.1.

By following Box 2.2, one can define as many variables as necessary. However, in many cases, more than thousands of decision variables are needed. In that case, it is not logical to write thousands of lines of code just to define the variables. To solve this issue, one must use sets. For instance, assume x is the amount of solar electricity one wants to sell to the grid. The person wants to find the optimal hourly value of x during a typical day; Therefore, the aim is to find the optimal x at hour 1, hour 2, ..., hour 24, and there are 24 decision variables. In this case one can define a set and use it as shown in Box 2.3 to define all 24 decision variables.

Box 2.3

```
time = [i for i in range(1,24 + 1)]
model.x = pyo.Var(time, within = pyo.PositiveReals, bounds
= (0,5))
```

Table 2.1 Possible domain values (pay attention to capital letters) [1]

Value	Definition
`Reals`	$(-\infty, +\infty)$
`PositiveReals`	$(0, +\infty)$
`NonPositiveReals`	$(-\infty, 0]$
`NegativeReals`	$(-\infty, 0)$
`NonNegativeReals`	$[0, +\infty)$
`PercentFraction OR UnitInterval`	[0,1]
`Integers`	$(-\infty, +\infty)$ only integers values
`PositiveIntegers`	$(0, +\infty)$ only integers values
`NonPositiveIntegers`	$(-\infty, 0]$ only integers values
`NegativeIntegers`	$(-\infty, 0)$ only integers values
`NonNegativeIntegers`	$[0, +\infty)$ only integers values
`Boolean OR Binary`	{0,1}

Line 1 defines the set of time using a `for` loop and range function. `range(a,b)` generates numbers starting from a and ending at b-1; Therefore, to include be in the generated numbers, one must use `range(a,b + 1)`. Line 2 of Box 2.3 defines 24 variables $x_1, x_2, \ldots, x_{24}$. If one wishes to define a variable using more than one set, they can add another list as the argument. For instance, `pyo.Var(time, time, ...)`, defines variables as $x_{1,1}, x_{1,2}, \ldots, x_{24,24}$. Note that when defining a variable using a set, the value of `within` and `bounds` will be applied to all 24 variables. Similarly, if one assigns a specific value for initialization using the `initialize` argument, that value will be used for all 24 variables. Now, the question is how to change the upper and lower bounds for one or more specific variables (e.g., x_{13}). To do so, one can use a *rule function* as indicated in Box 2.4.

Box 2.4

```
1 def bound_rule(model, t):
2    if t = = 13:
3        return (1,4)
4    else:
5        return (0,5)
6 model.x = pyo.Var(time, within = pyo.PositiveReals, bounds
  = bound_rule)
```

In this code, the lower and upper bound of all variables except x_{13} is (0,5). The same approach can be used to change the initial value of a specific variable. The rule functions (lines 1–5 in Box 2.4) are particularly important when defining constraints since it is easier for the modeler to define the constraint using rule functions instead of expressions. Note that *model* is always one of the inputs of the rule function (see line 1). In addition, if we want to use the rule function over a set, an iterator like t should also be an input where $t \in time$. After defining the decision variables, other parts of the optimization problem such as the objective function and constraints can be defined.

2.2.2 *Objective Function*

The objective function is the function that should be minimized or maximized. To define an objective function in Pyomo one must use `pyo.Objective()` and add the objective function to the model as shown in Box 2.5.

Box 2.5

```
1 import pyomo.environ as pyo
2 model = pyo.ConcreteModel()
3 model.x1 = pyo.Var(within = pyo.NonNegativeReals)
4 model.x2 = pyo.Var(within = pyo.NonNegativeReals)
5 model.obj = pyo.Objective(expr = 2*model.x1 + 4*model.x2, sense = pyo.maximize)
```

Line 5 in Box 2.5 defines the objective function called `obj` which should be maximized since `sense` is equal to `pyo.maximize`. The default value for `sense` is `pyo.minimize`. If the number of decision variables is more, `sense` cannot be used. In this case, rule functions that were introduced in the previous section should be utilized as shown in Box 2.6.

Box 2.6

```
N = [1, 2, 3, 4]
model.x = pyo.Var(N, within = pyo.NonNegativeReals, bounds = (0,3))
c = {1:2, 2:6, 3:9, 4:0.5}
def obj_rule(model):
    return sum(c[i]*model.x[i] for i in N)
model.obj = pyo.Objective(rule = obj_rule, sense = pyo.maximize)
```

In the code presented in Box 2.6, there are four decision variables as indicated by the set N. The cost of choosing each decision variable is determined by a dictionary called c, meaning that the objective function is $\sum_N c_i x_i$. c is a parameter for the optimization problem and its values are constant (the cost of choosing the first variable is 2, that of the second variable is 6, and so on). If one wants to write this objective as an expression, it would be considerably longer: $c_1x_1+c_2x_2+c_3x_3+c_4x_4 = 2x_1+6_{x2}+9_{x3}+0.5x_4$. Note that instead of using a dictionary, a list can also be used to define c (replace line 3 with `c = [2,6,9,0.5]`); In this case, `c[i]` in line 5 should be changed to `c[i-1]` since the list index in Python starts from 0 and Pyomo sets start from 1. To check whether the objective function is defined correctly, one can use `print(model.obj.expr)` which returns `2*x[1] + 6*x[2] + 9*x[3] + 0.5*x[4]`. Using a rule function, any objective function no matter how complex can easily be modeled in Pyomo. The next section describes the constraints and how to add them to the optimization problem in Pyomo.

2.2.3 *Constraints*

Constraints are one of the most important parts of an optimization problem. They define a feasible region in which the optimizer should search to find the optimal values of the decision variables in such a way that maximizes or minimizes the objective function. If such a common region does not exist, the problem becomes infeasible where there is no feasible region to be searched. In other words, the values of decision variables should not only maximize or minimize the objective function, but also satisfy the constraints. To define a constraint in Pyomo, the steps described in Box 2.7 should be taken.

Box 2.7

```
model.con = pyo.Constraint(expr = expression of the
constraints)
```

As an example, the expression could be `4*model. x 1 + 2*model. x 2 >=1`. Three types of constraints are equality (==), greater than or equal (≥), and less than or equal (≤). Note that from the mathematical perspective, there is no difference between ≥ and > or between ≤ and < if the variable is continuous. Now, assume that the following constraints (2.5)–(2.7) are given and they should be translated into Pyomo.

$$x_1 + x_1 \leq 2 \tag{2.5}$$

$$x_1 - x_2 \leq 3 \tag{2.6}$$

$$x_1 - x_2 \geq 5 => -x_1 + x_2 \leq -5 \tag{2.7}$$

As can be seen, there are three constraints and two variables (the third constraint is multiplied by -1 so that all constraints are in the form of less than or equal). These constraints can be translated into a matrix format as shown by Eq. (2.8).

$$\begin{bmatrix} 1 & 1 \\ 1 & -1 \\ -1 & 1 \end{bmatrix} \times \begin{bmatrix} x_1 \\ x_2 \end{bmatrix} \leq \begin{bmatrix} 2 \\ 3 \\ -5 \end{bmatrix} \tag{2.8}$$

Having the first matrix $\left(A = \begin{bmatrix} 1 & 1 \\ 1 & -1 \\ -1 & 1 \end{bmatrix} \right)$ and second matrix $\left(B = \begin{bmatrix} 2 \\ 3 \\ -5 \end{bmatrix} \right)$ gives us enough information to model the problem as shown in Box 2.8.

Box 2.8

```
1 V = [1, 2]     # set of decision variables
2 C = [1, 2, 3]   # set of constraints
3 A = {(1,1):1, (1,2):1,
4       (2,1):1, (2,2):-1,
5       (3,1):-1, (3,2):1}
6 B = {1:2, 2:3, 3:-5}
7 def con_rule(model,c):
8     return sum(A[c,v]*model.x[v] for v in V) < = B[c]
9 model.con = pyo.Constraint(C, rule = con_rule)

10 opt = pyo.SolverFactory("solver name")
11 results = opt.solve(model)
12 print(pyo.check_optimal_termination(results))
```

Box 2.8 shows that there are a set of decision variables, V, and a set of constraints, C. Matrix A is defined using a dictionary in which keys are defined as (c, v) where $c \in C, v \in V$, meaning that the first value in the tuple is the index of the constraint and the second value is the index of the variable. Similarly, matrix B is defined in which the keys are constraint indices. Finally, constraints are defined over C (similar to defining a variable over a set, in Box 2.3). Iterator c is an input to the rule function for the same reason described below the Box 2.4. The rule function returns true or false values, meaning that if for a certain set of values, the left-hand side is less than or equal to the right-hand side, it returns true, otherwise, it is false. After defining the optimization problem using the above codes, one can solve it using an optimizer. The optimizer's name is specified in line 10, then the model is solved in line 11 and the optimality is checked in line 12. If the output of line 12 is "True", the problem is solved properly and global optimal values are found by the optimizer.

Now that we have all the basic information on how to formulate an optimization problem in Pyomo, it is time to solve the example problems to consolidate the gained knowledge.

2.3 Example Problems

This section provides six example problems categorized into four groups. For each example, the Pyomo code along with the results are presented.

2.3.1 *Shortest Path*

The aim of solving the shortest path problem (SPP), as the name implies, is to find the path with the minimum distance between a starting and an ending point. Assume the starting point on the map shown in Fig. 2.1 is node one and the destination is node seven. The numbers on the edges are the distance between two nodes which should be minimized.

We start by developing the mathematical formulation of the problem and then we translate it into Pyomo. The decision variables of the problem are defined as x_{ij} such that i and j are the node number and x is a binary variable. Therefore, x_{12} refers to the path between node 1 and node 2. If $x_{12} = 1$, the path is chosen as a part of the optimal route, otherwise ($x_{12} = 0$), it is not included in the optimal route. The objective function (OF) is shown in Eq. (2.9).

$$\begin{aligned} \min OF = {} & 5x_{12} + 3x_{13} + 4x_{23} + x_{24} + 7x_{25} \\ & + 6x_{32} + 5x_{34} + x_{35} + 2x_{36} + 9x_{45} + 4x_{47} \end{aligned}$$

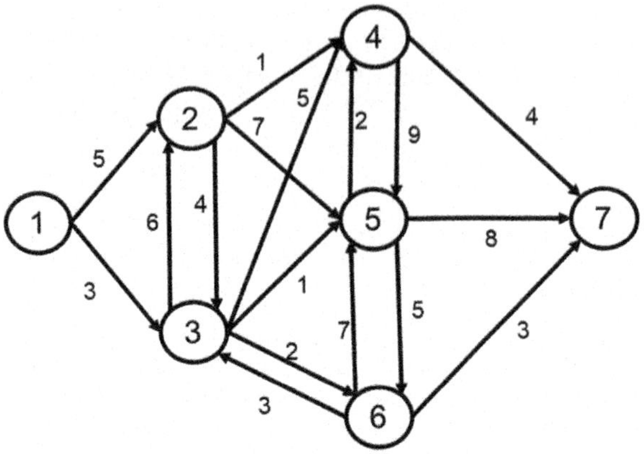

Fig. 2.1 Map of the shortest path problem

$$+ 2x_{54} + 5x_{56} + 8x_{57} + 3x_{63} + 7x_{65} + 3x_{67}, \tag{2.9}$$

where the constant multipliers are the distances shown by Fig. 2.1. If we solve the problem without the constraints, the optimum value of the OF would be zero ($x_{ij} = 0$ for all i, j) since we do not force the problem to provide a route from 1 to 7. Equations (2.10)–(2.16) show the constraints of the problem.

$$1 = x_{12} + x_{13} \tag{2.10}$$

$$x_{12} + x_{32} = x_{23} + x_{24} + x_{25} \tag{2.11}$$

$$x_{13} + x_{23} + x_{63} = x_{32} + x_{34} + x_{35} + x_{36} \tag{2.12}$$

$$x_{24} + x_{34} + x_{54} = x_{45} + x_{47} \tag{2.13}$$

$$x_{25} + x_{35} + x_{45} + x_{65} = x_{54} + x_{56} + x_{57} \tag{2.14}$$

$$x_{36} + x_{56} = x_{63} + x_{65} + x_{67} \tag{2.15}$$

$$x_{47} + x_{57} + x_{67} = 1 \tag{2.16}$$

The constraints are input–output balance (left side is input and right side is output) over every node. For instance, Eq. (2.11) shows that the sum of inputs to node two must be equal to the sum of the outputs from node two. Since the starting point is node one, the input to this node is forced to be 1 as shown in Eq. (2.10). Similarly, since the destination is node 7, its output is forced to be one as presented in Eq. (2.16). After defining the decision variables, objective function, and constraints, it is time to solve the problem, but before that, the problem should be formulated in Pyomo. Decision variables are defined in Pyomo based on Box 2.9.

Box 2.9

```
import pyomo.environ as pyo
model = pyo.ConcreteModel(name = "Shortest Path")

model. x 12 = pyo.Var(within = pyo.Binary)
model. x 13 = pyo.Var(within = pyo.Binary)

model. x 23 = pyo.Var(within = pyo.Binary)
model. x 24 = pyo.Var(within = pyo.Binary)
model. x 25 = pyo.Var(within = pyo.Binary)
```

```
model. x 32 = pyo.Var(within = pyo.Binary)
model. x 34 = pyo.Var(within = pyo.Binary)
model. x 35 = pyo.Var(within = pyo.Binary)
model. x 36 = pyo.Var(within = pyo.Binary)

model. x 45 = pyo.Var(within = pyo.Binary)
model. x 47 = pyo.Var(within = pyo.Binary)

model. x 54 = pyo.Var(within = pyo.Binary)
model. x 56 = pyo.Var(within = pyo.Binary)
model. x 57 = pyo.Var(within = pyo.Binary)

model. x 63 = pyo.Var(within = pyo.Binary)
model. x 65 = pyo.Var(within = pyo.Binary)
model. x 67 = pyo.Var(within = pyo.Binary)
```

Note that the formulation presented in Box 2.9 is introductory and in the second half of this section, a more sophisticated formulation will be presented. To define the OF and constraints, the code presented in Box 2.10 should be used.

Box 2.10

```
model.obj = pyo.Objective(expr = 5*model. x 12 + 3*model. x 13
+ 4*model. x 23 + 1*model. x 24 + 7*model. x 25 + 6*model. x 32
+ 5*model. x 34 + 1*model. x 35 + 2*model. x 36 + 9*model. x
45 + 4*model. x 47 + 2*model. x 54 + 5*model. x 56 + 8*model.
x 57 + 3*model. x 63 + 7*model. x 65 + 3*model. x 67, sense =
pyo.minimize)

model.N1 = pyo.Constraint(expr = 1 = = model. x 12 + model. x
13)

model.N2 = pyo.Constraint(expr = model. x 12 + model. x 32 =
= model. x 23 + model. x 24 + model. x 25)

model.N3 = pyo.Constraint(expr = model. x 13 + model. x 23 +
model. x 63 = = model. x 32 + model. x 34 + model. x 35 + model.
x 36)

model.N4 = pyo.Constraint(expr = model. x 24 + model. x 34 +
model. x 54 = = model. x 45 + model. x 47)

model.N5 = pyo.Constraint(expr = model. x 25 + model. x 35 +
model. x 45 + model. x 65 = = model. x 54 + model. x 56 + model.
x 57)
```

```
model.N6 = pyo.Constraint(expr = model. x 36 + model. x 56 =
= model. x 63 + model. x 65 + model. x 67)

model.N7 = pyo.Constraint(expr = model. x 47 + model. x 57 +
model. x 67 = = 1)
```

In Box 2.10, `model.obj` is the objective function which is defined using an expression following the mathematical formulation presented in Eq. (2.5). The constraints of every node are brought after the objective function following Eqs. (2.6)–(2.12). To solve the problem, GLPK solver is used as shown in Box 2.11. First, the optimizer is defined in line 1 and then the results are obtained in line 2. Finally, to check whether the solution is optimal and the problem is not infeasible or unbounded, line 3 is added. If the output of line 3 is True, then the problem is optimal and the solver successfully reached the optimal solution.

Box 2.11

```
1 opt = pyo.SolverFactory("glpk")
2 results = opt.solve(model)
3 print(pyo.check_optimal_termination(results))
```

The optimal values of the decision variables are listed in Table 2.2.

Table 2.2 shows that the optimal route is 1 → 3 → 6 → 7 and the optimal OF value (i.e. minimum distance) is 8. While the above formulation solves the problem and provides and optimal answer (since the output of Box 2.11, line 3 is True), the formulation is too long and if the number of nodes and edges increases, one cannot use this method. In the following, a better formulation using sets is provided.

Suppose the starting point on the map shown in Fig. 2.2 is one and the end point is ten. The distances between nodes are provided in Table 2.3.

If there is a path between two nodes, the distance is brought in Table 2.3, otherwise, the distance is set to a large number. Doing so ensures that the optimizer will not pick

Table 2.2 Decision variables' optimal value

Variable	Optimal value	Variable	Optimal value	Variable	Optimal value
x_{12}	0	x_{34}	0	x_{56}	0
x_{13}	1	x_{35}	0	x_{57}	0
x_{23}	0	x_{36}	1	x_{63}	0
x_{24}	0	x_{45}	0	x_{65}	0
x_{25}	0	x_{47}	0	x_{67}	1
x_{32}	0	x_{54}	0		

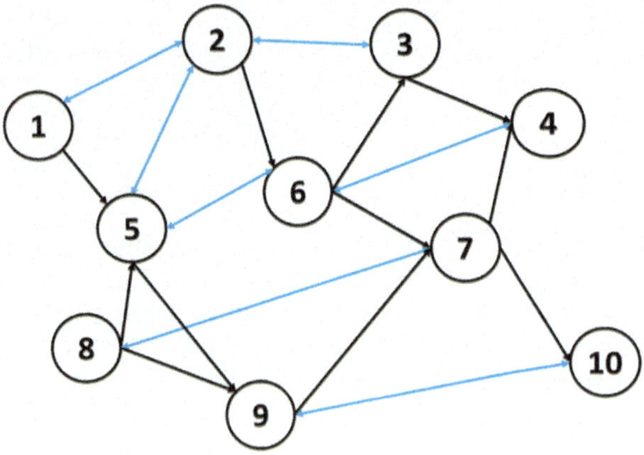

Fig. 2.2 Map of the shortest path problem (blue edges are bidirectional and black edges are unidirectional)

that path. For instance, there is no path from a node to itself; Therefore, the numbers on the main diagonal are 1,000,000. The mathematical formulation of the problem is presented in Eqs. (2.17)–(2.21).

$$\min OF = \sum_{R,C} D_{r,c} \times x_{r,c} \tag{2.17}$$

$$\sum_{R} x_{c,r} = 1 \qquad c = start \tag{2.18}$$

$$\sum_{R} x_{r,c} = 1 \qquad c = end \tag{2.19}$$

$$\sum_{R} x_{c,r} = \sum_{R} x_{r,c} \qquad \forall c \in C \tag{2.20}$$

$$x_{r,c} \in \{0,1\} \tag{2.21}$$

These equations show that, in general, this type of problem can be formulated using only three constraints [Eqs. (2.18)–(2.20)]. R and C are sets of nodes (see Table 2.3) and r and c represent a member in these sets. *Start* is the arbitrary starting node which must be a number between 1 and 10. Similarly, the *end* is the destination node. Equation (2.18) ensures that there is at least one output from the starting point while Eq. (2.19) ensures that there is at least one input to the destination node. Equation (2.20) states that if a node has an input, it must also have an output (balance). To formulate this problem in Pyomo, we follow the code presented in Box 2.12.

Box 2.12

Table 2.3 Distance between every two nodes

		Node number (C)									
		1	2	3	4	5	6	7	8	9	10
Node number (R)	1	1,000,000	**23**	1,000,000	1,000,000	**20**	1,000,000	1,000,000	1,000,000	1,000,000	1,000,000
	2	**23**	1,000,000	**91**	1,000,000	**54**	**11**	1,000,000	1,000,000	1,000,000	1,000,000
	3	1,000,000	**91**	1,000,000	**59**	1,000,000	1,000,000	1,000,000	1,000,000	1,000,000	1,000,000
	4	1,000,000	1,000,000	1,000,000	1,000,000	1,000,000	**34**	1,000,000	1,000,000	1,000,000	1,000,000
	5	1,000,000	**54**	1,000,000	1,000,000	1,000,000	**19**	1,000,000	1,000,000	**20**	1,000,000
	6	1,000,000	1,000,000	**54**	**34**	**19**	1,000,000	**92**	1,000,000	1,000,000	1,000,000
	7	1,000,000	1,000,000	1,000,000	**83**	1,000,000	1,000,000	1,000,000	**43**	1,000,000	**86**
	8	1,000,000	1,000,000	1,000,000	1,000,000	**72**	1,000,000	**43**	1,000,000	**50**	1,000,000
	9	1,000,000	1,000,000	1,000,000	1,000,000	1,000,000	1,000,000	**78**	1,000,000	1,000,000	**49**
	10	1,000,000	1,000,000	1,000,000	1,000,000	1,000,000	1,000,000	1,000,000	1,000,000	**49**	1,000,000

```
import pandas as pd
import pyomo.environ as pyo
D = pd.read_excel("excel_file_name.xlsx", header = 0,
index_col = 0)
C = list(D.columns)
R = list(D.index)
start = 1
end = 10
model = pyo.ConcreteModel(name="Shortest_Path")
model.x = pyo.Var(R, C, within=pyo.Binary)
def obj_rule(model):
    return sum(D.at[r,c]*model.x[r,c] for r in R for c in C)
model.obj = pyo.Objective(rule = obj_rule)
def balance_rule(model,c):
    if c == start:
        return sum( model.x[c,r] for r in R ) == 1
    elif c == end:
        return sum( model.x[r,c] for r in R ) == 1
    else:
        return sum( model.x[r,c] for r in R ) == sum( model.x[c,r]
for r in R )
model.balance = pyo.Constraint(C, rule = balance_rule)
res = pyo.SolverFactory("glpk").solve(model)
print(pyo.check_optimal_termination(res))
print(f"Objective value = {pyo.value(model.obj)}")
print("Optimal Path:")
for r in R:
    for c in C:
        if pyo.value(model.x[r,c]) == 1:
            print(f"{r} => {c}", end=" ")
```

According to Box 2.12, we first read the distance matrix (Table 2.3) from Excel using the Pandas library (see Pandas documentation for more information). Then, we define R and C sets and specify start and end points. After creating the model and defining the binary decision variables, we define the OF using a rule function. Since the objective is to minimize the total distance, the expression `sum(D.at[r,c]*model.x[r,c] for r in R for c in C)` is used which is equivalent to $\sum_{R,C} D_{r,c} \times x_{r,c}$. Moreover, since the default sense of Pyomo is minimization, `sense = pyo.minimize` is not specified in the OF. Lines 13–20 define the constraints using a single rule function and `if` statements for start, destinations, and other nodes. Finally, the problem is solved using GLPK in line 21 and an optimality check is conducted in line 22. To attain the optimal OF value, `pyo.value(model.obj)` is used and to get the optimal values of the decision variables, `model.obj` is replaced with `model.x[r,c]`. The optimum path is $1 \rightarrow 5 \rightarrow 9 \rightarrow 10$.

Note that in both versions of this problem, we are sure that there is at least one route from the start to the destination. Otherwise, the problem would be infeasible. In the next section, another two problems known as hub/facility/warehouse location allocation are discussed.

2.3.2 *Warehouse Location*

Warehouse location allocation (WLA) is a linear programming problem in which the aim is to find the optimum places, from a set of given locations, to build a warehouse or facility such that the distance of the warehouse from its neighbor nodes is minimized. This problem can be formulated in different forms. Here, the first formulation is based on a limitation on the number of connections between the warehouses and neighbors, and the second is based on a limitation on the warehouse coverage radius. To formulate the problem, we need a set including the nodes such as $N = \{1, 2, 3, \ldots, N_{max}\}$, and two sets of decision variables as shown in Eqs. (2.18) and (2.19). It is assumed that n is the warehouse index, and n' is city index.

$$y_n = \begin{cases} 1 & \text{if there is a warehouse in node } n \\ 0 & \text{if there is not a warehouse in node } n \end{cases} \quad n \in N \tag{2.22}$$

$$x_{n,nn} = \begin{cases} 1 & \text{if there is a connection between warehouse } n \text{ and city } n' \\ 0 & o.w. \end{cases} \quad n\&n' \in N \tag{2.23}$$

The OF and constraints are represented in Eqs. (2.24)–(2.27).

$$\min OF = \sum_{n\&n' \in N,\ n \neq n'} D_{n,n'} \times x_{n,n'} \tag{2.24}$$

$$\sum_{n \in N} y_n \leq W_{\max} \tag{2.25}$$

$$\sum_{n' \in N} x_{n,n'} \leq y_n \times (C_{max} + 1) \quad \forall n \in N \tag{2.26}$$

$$\sum_{n \in N} x_{n,n'} = 1 \qquad \forall n' \in N \tag{2.27}$$

The OF is simply the sum of the distances between all facilities n and all cities n' such that $n \neq n'$. Equation (2.25) states that the total number of warehouses (left side) must be less than or equal to the allowable number of warehouses (W_{max}). Equation (2.26) states that the maximum connections of warehouse n (if $y_n = 1$) are limited by the maximum number of connections (C_{max}). Since node n itself is considered in the equation, a + 1 is added to the C_{max}. It means, excluding node n itself, the node can have C_{max} connections. Finally, constraint (2.27) ensures that every city n' is connected to a warehouse n. The problem is feasible only when $W_{max} \times C_{max} \geq N_{max}$. Box 2.13 shows the Pyomo formulation of the problem. In this code, nn is equivalent to n' in the mathematical form.

Box 2.13

```
num = 100
max_w = 10
max_c = 15
model = pyo.ConcreteModel(name="Warehouse location")
N = list(range(1, num +1))
model.y = pyo.Var(N, within=pyo.Binary)
model.x = pyo.Var(N, N, within=pyo.Binary)

def obj_rule(model):
    return sum(D[n,nn]*model.x[n,nn] for n in N for nn in N if n!=nn)
model.obj = pyo.Objective(rule=obj_rule)

def max_w_rule(model):
    return sum(model.y[n] for n in N) <= max_w
model.max_w = pyo.Constraint(rule=max_w_rule)

def w_city_connection_rule(model,n):
    return sum(model.x[n,nn] for nn in N) <= model.y[n]*(max_c +1)
model.w_city_connection= pyo.Constraint(N, rule=w_city_connection_rule)

def connection_rule(model,nn):
    return sum(model.x[n,nn] for n in N) == 1
model.connection = pyo.Constraint(N, rule=connection_rule)
```

The first three lines define the parameters of the optimization problem. Also, `num` is the number of nodes, `max_w` is the maximum number of warehouses, and `max_c` is the maximum number of connections for a facility. After defining the set of nodes in line 4, y_n and $x_{n,n'}$ are defined in lines 5 and 6. OF formulation is the same as Eq. (2.20). Lines 10–12 show Eq. (2.21), lines 13–15 show Eq. (2.22), and lines 16–18 show Eq. (2.23). The distance matrix D can be defined similar to the SPP. Figure 2.3a, b illustrate a sample set of nodes and the optimal location of warehouses, respectively.

In order to increase the complexity of the problem, we can impose a restriction on the coverage radius of warehouses rather than restricting the number of connections. In addition, construction costs can be considered for each warehouse, meaning that we should pay a price to establish a warehouse in a node. Set and decision variables are the same as before and the mathematical formulation is presented in Eqs. (2.28)–(2.31).

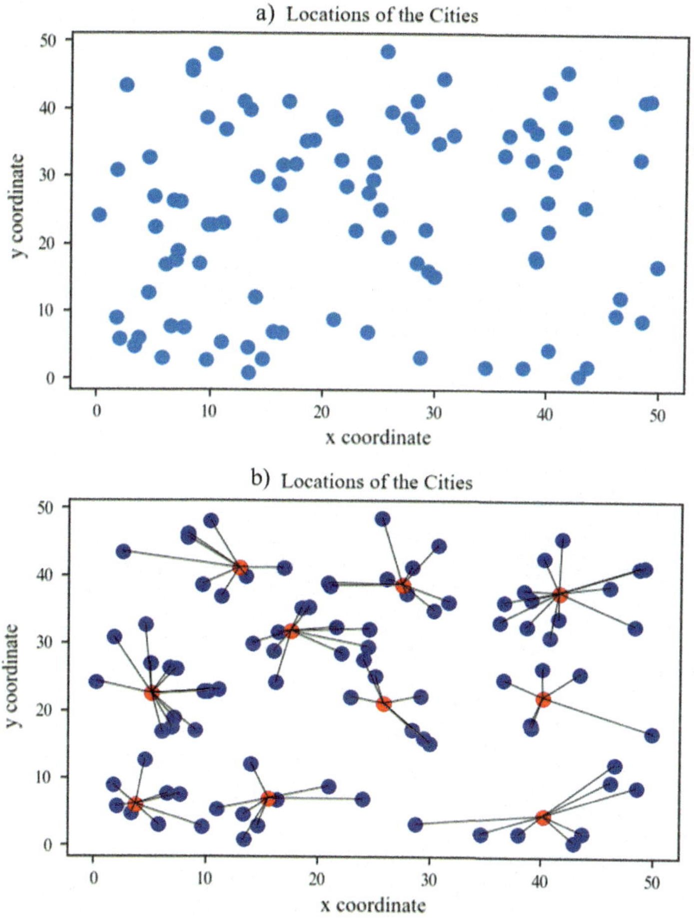

Fig. 2.3 **a** A sample distribution of nodes in the x–y coordinate, **b** Optimal location of warehouses (red) and their connected cities

$$\min OF = \sum_{n\&n' \in N,\ n \neq n'} D_{n,n'} \times x_{n,n'} + CC \sum_{n \in N} y_n \tag{2.28}$$

$$\sum_{n \in N} y_n \leq W_{\max} \tag{2.29}$$

$$D_{n,n'} \times x_{n,n'} \leq y_n \times R_{\max} \quad \forall n \ \& \ n' \in N \tag{2.30}$$

$$\sum_{n \in N} x_{n,n'} = 1 - y_{n'} \quad \forall n' \in N, n \neq n' \tag{2.31}$$

The second term of the OF is added to take into account the impact of construction costs (CC). Equation (2.29) is the same as Eq. (2.25). Equation (2.30) ensures that if node n is a warehouse, the distance of every connected city is less than or equal to the maximum allowable radius. Equation (2.31) mentions that each city, n', must be connected to a warehouse, n, if the city itself is not a warehouse ($y_{n'} = 0$). Pyomo code of these equations is brought in Box 2.14.

Box 2.14

```
Num_city = 100
max_facility = 20
max_radious = 0.5*max_dist
cost_facility = max_dist

model = pyo.ConcreteModel(name="Facility location2")

N = list(range(1, Num_city +1))
model.y = pyo.Var(N, within=pyo.Binary)
model.x = pyo.Var(N, N, within=pyo.Binary)

def obj_rule(model):
    return sum(D[n,nn]*model.x[n,nn] for n in N for nn in N if n!=nn) + sum(model.y[n]*cost_facility for n in N)
model.obj = pyo.Objective(rule=obj_rule)

def max_facility_rule(model):
    return sum(model.y[n] for n in N) <= max_facility
model.max_facility = pyo.Constraint(rule=max_facility_rule)

def connection_rule(model,nn):
    return sum(model.x[n,nn] for n in N if n!=nn) == 1-model.y[nn]
model.connection = pyo.Constraint(N, rule=connection_rule)
```

```
18 def radious_rule(model,n,nn):
19     if n != nn:
20         return D[n,nn]*model.x[n,nn] <= model.y[n]*max_radious
21     else:
22         return pyo.Constraint.Skip
23 model.radious = pyo.Constraint(N,N, rule=radious_rule)
```

The first four lines are the parameters of the problem; Maximum radius is set to be half of the maximum distance between two nodes, and the cost of building a warehouse is equal to the maximum distance between two nodes. If the cost is too high, the model uses the minimum number of warehouses and if the cost is too low, the model builds all 20 warehouses. The updated version of the OF based on Eq. (2.28) is brought in lines 9–11. Lines 12–14 are based on Eq. (2.29), lines 15–17 are the same as Eq. (2.31), and lines 18–23 are based on Eq. (2.30). The command `pyo.Constraint.Skip` can be used when there is no constraint for a specific condition. For instance, in the above code, if the index of the warehouse and city are the same, there is no need to have a constraint. Skipping reduces the number of constraints and lowers the problem run time. Figure 2.4a, b shows a sample distribution of nodes and the optimal location of warehouses.

As illustrated by Fig. 2.4, seven nodes (7, 13, 17, 52, 60, 89, 98) are chosen as the optimal warehouse locations. The optimization result is brought in Table 2.4.

Note that in this sample problem, the maximum allowable radius is 60.75 and as shown in Table 2.4, the maximum distance covered by a warehouse is less than this number. In the next section, one of the most famous optimization problems called the traveling salesperson (TSP) is discussed and solved using Pyomo.

2.3.3 *Travelling Salesman*

TSP is a famous optimization problem in the world of linear programming. The question is, if a list of cities and the distances between them is available, what's the shortest route that goes through each city only once and comes back to the starting city? It is similar to the SPP but in SPP the goal is to find the shortest path from a starting point to a destination and in TSP the goal is to start from a node (no matter which one), visit every available node, and come back to the starting point in a way that the total covered distance is minimized. For illustration, consider Fig. 2.5.

In this map, there are four cities and the starting node is one. If we follow the route shown in Fig. 2.5a the total distance will be 4 while following Fig. 2.5b path will result in a total distance of around 4.83. Therefore, the optimal route is (a). Mathematical formulation of the problem is brought in Eqs. (2.32)–(2.36).

$$\min OF = \sum_{i \in I, o \in O, i \neq o} D_{i,o} \times x_{i,o} \tag{2.32}$$

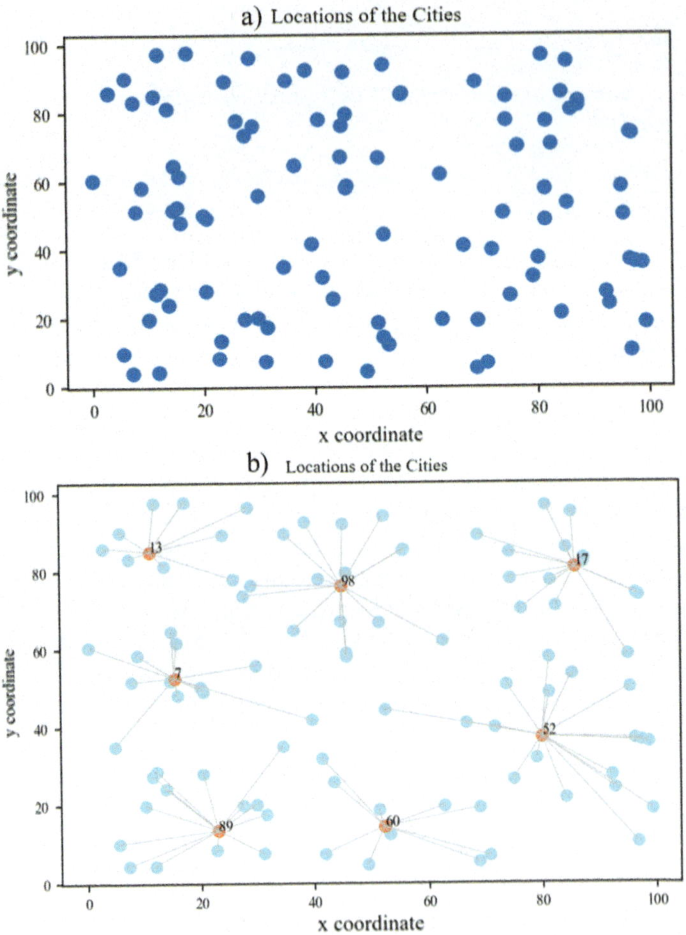

Fig. 2.4 **a** A sample distribution of nodes in the x–y coordinate, **b** Optimal location of warehouses (orange) and their connected cities

Table 2.4 Optimization results for the WLA problem with radius constraint

Warehouse node	Number of connections	Maximum distance	Minimum distance
7	12	26.49	1.1
13	9	20.75	3.99
17	14	24.23	1.71
52	18	31.8	5.53
60	10	20.69	2.21
89	14	24.4	4.95
98	16	22.56	3.44

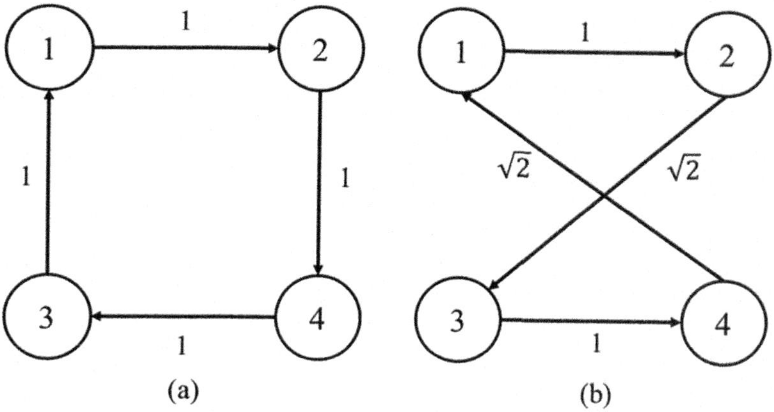

Fig. 2.5 **a** A route with minimum distance covered, **b** a suboptimal route

$$\sum_{i \in I, i \neq o} x_{i,o} = 1 \quad \forall o \in O \tag{2.33}$$

$$\sum_{o \in O, i \neq o} x_{i,o} = 1 \quad \forall i \in I \tag{2.34}$$

$$u_i - u_o + N \times x_{i,o} \leq N - 1 \quad 2 \leq i \neq o \leq N \quad \forall i \in I \;\&\; o \in O \tag{2.35}$$

$$x_{i,o} = \begin{cases} 1 \text{ if there is a connection between node } i \text{ and } o \\ 0 \qquad\qquad o.w. \end{cases} \quad i \in I \text{ and } o \in O$$

$$u_i \;\&\; u_o \in \mathbb{N} \tag{2.36}$$

where I I is the set of inputs and O is the set of outputs, and $I = O = \{1,2,3,\ldots,N\}$ where N is the number of cities. The OF is the sum of the distances between two nodes or the total covered distance. Equation (2.33) mentions that for every node o, there should be exactly one input from other nodes in set I, such that $i \neq o$ (i.e., a node cannot be connected to itself). Equation (2.34) states that for every node i there should be exactly one output to the other nodes in set O, such that $i \neq o$. Constraint (2.35) is known as Miller-Tucker-Zemlin (MTZ) constraint and ensures the formation of a single tour instead of multiple subtours. To clarify, consider Fig. 2.6 in which two solutions for the map are provided.

Since the second solution has a lower total distance, the optimizer will choose this solution if MTZ is not considered. More information about MTZ constraint can be found in [3]. In order to formulate the TSP problem in Pyomo, the code presented in Box 2.15 should be taken into account.

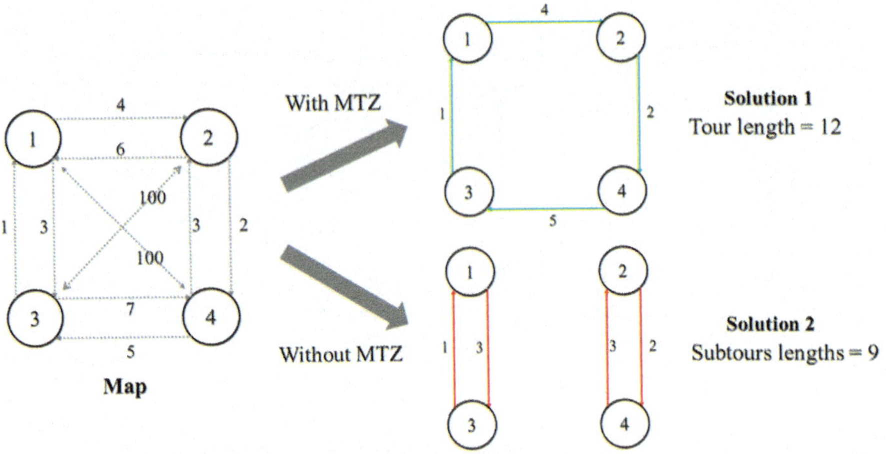

Fig. 2.6 The effect of MTZ constraint on the optimal result

Box 2.15

```
model = pyo.ConcreteModel(name="TSP")
model.x = pyo.Var(R, C, within=pyo.Binary)
model.u = pyo.Var(R, within=pyo.NonNegativeIntegers)
def obj_rule(model):
    return sum(Dist[r,c]*model.x[r,c] for r in R for c in C)
model.obj = pyo.Objective(rule=obj_rule)
def row_rule(model, r):
    return sum(model.x[r,c] for c in C) == 1
model.row = pyo.Constraint(R, rule=row_rule)
def col_rule(model, c):
    return sum(model.x[r,c] for r in R) == 1
model.col = pyo.Constraint(C, rule=col_rule)
def visit_rule(model,r,c):
    if r != c:
        return model.u[r] - model.u[c] + len(R)*model.x[r,c] <= len(R)-1
    else:
        return pyo.Constraint.Skip
model.visit=pyo.Constraint(list(set(R)-{1}),list(set(C)-{1}),rule=visit_rule)
```

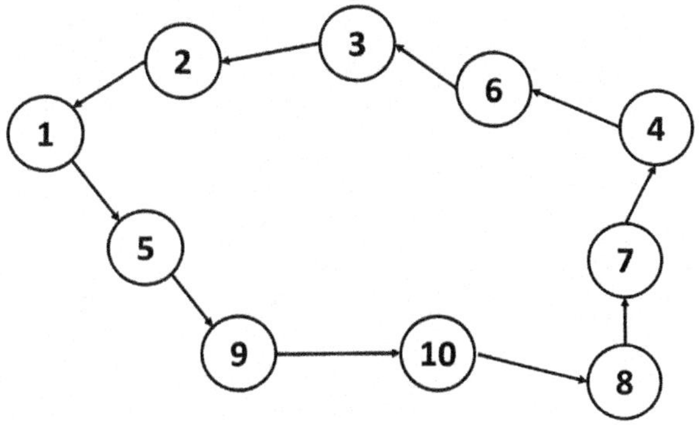

Fig. 2.7 The optimal tour for the TSP problem

The OF and the first two constraints are similar to the SPP, line 3 defines the integer variable u. The domain of u is `NonNegativeIntegers` since $u \in \mathbb{N}$. Lines 13–17 define the rule function of the MTZ constraint. When the indices of nodes are not equal, a constraint is defined using an `if` statement. Conversely, when the indices are the same, no constraint is applied. Note that R and C sets are the same as I and O in the mathematical formulation and `len(R)` is equal to N. The problem is solved using the nodes (without the paths) shown in Fig. 2.2 and the distances presented in Table 2.3. The optimal tour is depicted in Fig. 2.7.

The last problem discussed in this chapter is the N-queen problem. The next section is devoted to the explanation and Pyomo formulation of the N-queen problem which similar to TSP is a well-known optimization problem.

2.3.4 N-Queen

The aim of solving the N-queen problem is to find an optimal arrangement to place the maximum number of queens on a $N \times N$ chessboard so that neither one of them can attack the other. It turns out that for $N \geq 4$ ($N = 1$ is meaningless, and there is no answer for $N = 2, 3$), the maximum number of queens is equal to the chess board dimensions, meaning that one can place at most four queens on a four-by-four board in a way that they cannot attack each other. Two sample solutions for the 8-queen and 9-queen problems are depicted in Fig. 2.8.

To formulate this problem, we have to first define the sets and the decision variables. The sets are rows (R) and columns (C) of the board and $r \in R, c \in C$. The decision variable $x_{r,c}$ is a binary variable. If there is a queen in row, r, and column, c, then $x_{r,c} = 1$. Equations (2.37)–(2.43) show the mathematical formulation of the problem.

$$\max OF = \sum_{r \in R, c \in C} x_{r,c} \tag{2.37}$$

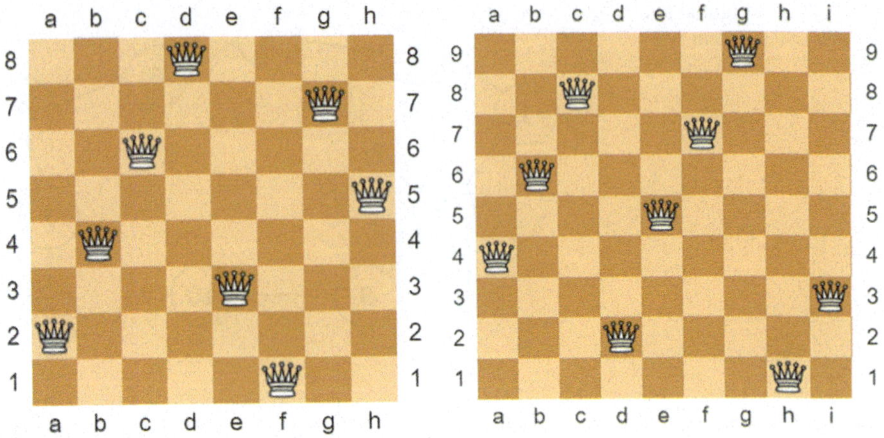

Fig. 2.8 Sample solutions for the 8-queen (left) and 9-queen (right) problems (Courtesy of Wikipedia)

$$\sum_{c \in C} x_{r,c} \leq 1 \quad \forall r \in R \tag{2.38}$$

$$\sum_{r \in R} x_{r,c} \leq 1 \quad \forall c \in C \tag{2.39}$$

$$x_{1,c} + \sum_{m=1}^{N-c} x_{1+m,c+m} \leq 1 \quad \forall c \in C \tag{2.40}$$

$$x_{r,1} + \sum_{m=1}^{N-r} x_{r+m,1+m} \leq 1 \quad \forall r \in R \tag{2.41}$$

$$x_{1,c} + \sum_{m=1}^{c-1} x_{1+m,c-m} \leq 1 \quad \forall c \in C \tag{2.42}$$

$$x_{r,N} + \sum_{m=1}^{N-r} x_{r+m,N-m} \leq 1 \quad \forall r \in R \tag{2.43}$$

Equation (2.37) is the OF which tries to maximize the number of queens on a board. Equations (2.38) and (2.39) mention that there should be at most one queen in a row and in a column, respectively. To consider the diagonal attack of the queens, we should add the remaining four equations. Equations (2.40) and (2.41) account for the forward diagonal attack which is shown in Fig. 2.9.

As can be seen in Fig. 2.9, forward diagonal attacks start from the first row ($x_{1,c}$) or the first column ($x_{r,1}$) and move towards one row down and one column forward (i.e., to the right). They continue moving until reaching the last column or

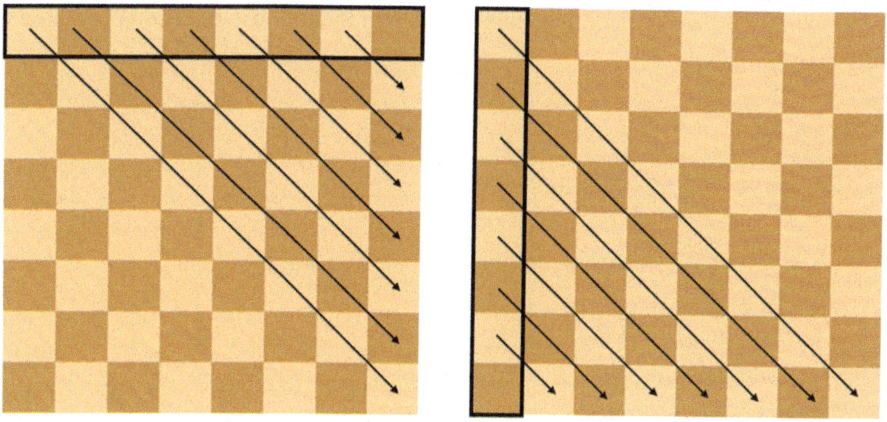

Fig. 2.9 Possible arrows for the forward diagonal attack from the first row (left) and the first column (right)

row. Equation (2.36) shows the forward diagonal movement from the first row while Eq. (2.41) shows the forward diagonal movement from the first column. In both cases, if r or $c = N$, the movement stops. Similarly, we can define constraints for the backward diagonals shown in Fig. 2.10.

Backward diagonal attacks start from either the first row or the last column. The logic is similar to the forward movement but here instead of moving to the right, we move to the left (backward). The movement stops when we reach the last row, or the first column. Equation (2.42) shows the backward move from the first row to the first column (stops when $c = 1$ and the upper bound of the summation becomes zero). Equation (2.43) shows the backward movement from the last column to the last row

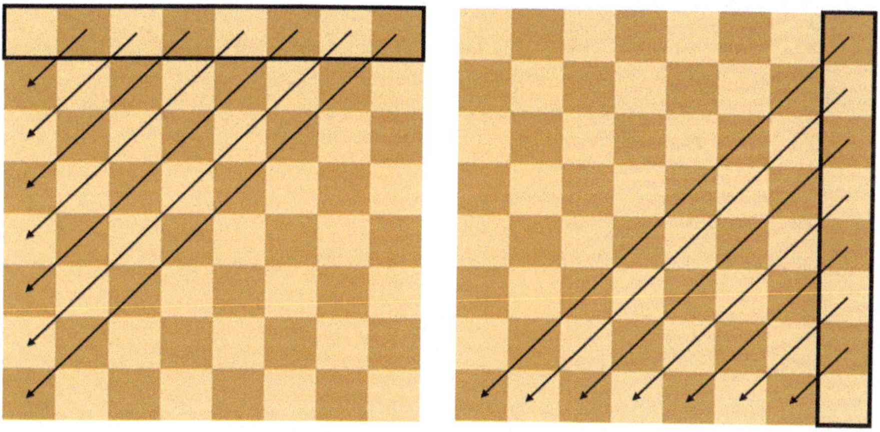

Fig. 2.10 Possible arrows for the backward diagonal attack from the first row (left) and the last column (right)

(stops when $r = N$). To formulate this problem in Pyomo, use the code presented in Box 2.16.

Box 2.16

```
import pyomo.environ as pyo
N = 20
model = pyo.ConcreteModel(name="NQueen")
R = list(range(1,N +1))
C = list(range(1,N +1))
model.x = pyo.Var(R, C, within=pyo.Binary)
def obj_rule(model):
    return sum(model.x[r,c] for r in R for c in C)
model.obj = pyo.Objective(rule=obj_rule, sense = pyo.maximize)
def one_in_row(model,r):
    return sum(model.x[r,c] for c in C) <= 1
model.one_in_row = pyo.Constraint(R, rule=one_in_row)
def one_in_col(model,c):
    return sum(model.x[r,c] for r in R) <= 1
model.one_in_col = pyo.Constraint(C, rule=one_in_col)
def forward_diag_top_row(model,c):
    return model.x[1,c]+sum(model.x[1+m,c+m] for m in list(range(1,N-c +1))) <= 1
model.forward_diag_top_row = pyo.Constraint(C, rule=forward_diag_top_row)
def forward_diag_first_col(model,r):
    return model.x[r,1]+sum(model.x[r+m,1+m] for m in list(range(1,N-r +1))) <= 1
model.forward_diag_first_col = pyo.Constraint(R, rule=forward_diag_first_col)
def backward_diag_top_row(model,c):
    return model.x[1,c]+sum(model.x[1+m,c-m] for m in list(range(1,c-1 +1))) <= 1
model.backward_diag_top_row = pyo.Constraint(C, rule=backward_diag_top_row)
def backward_diag_last_col(model,r):
```

```
        return model.x[r,N]sum( model.x[r+m,N-m] for m in list(range(1,N-r +1))) <= 1
    model.backward_diag_last_col = pyo.Constraint(R, rule=backward_diag_last_col)

solver = pyo.SolverFactory("highs")
res = solver.solve(model)
print(pyo.check_optimal_termination(res))

board = []
for i in range(N):
    for j in range(N):
        if pyo.value(model.x[i+1,j+1]) == 1:
            print('Q',end=' ')
        else:
            print('*',end=' ')
    print('')
```

In the above code, we do not import data from an external source like Excel. The definition of constraints and the OF is exactly similar to the mathematical formulation. Three points should be emphasized here; First, we can define an iterator in the rule function similar to `m`. Second, naming constraints and their rule function is important in readability, in the above code, we chose names that are relevant and help the reader find out what the constraint is about. Third, the upper bound of the `range` function when defining sets and in the constraints has a +1 which as mentioned before is because Python does not consider the upper bound itself. Finally, instead of using GLPK, HiGHS is used to solve the problem. GLPK works fine and can solve the problem but the aim is to show that other open-source solvers can be used. In chapter four, we mostly use HiGHS as it is faster than GLPK. You can download HiGHS from [4]. If you run the above code, you will see a solution like the one shown in Fig. 2.11.

Note that the N-queen problem could have multiple solutions and the above solution is one of the possible optimal solutions.

2.4 Concluding Remarks

The aim of this chapter is to provide the essential background for the reader so that they can understand different parts of an optimization problem, how to develop the mathematical formulation of an optimization problem, and how to translate the mathematical formulation to the Python programming language using the Pyomo library.

```
*  *  *  *  Q  *  *  *  *  *  *  *  *  *  *  *  *  *  *  *
Q  *  *  *  *  *  *  *  *  *  *  *  *  *  *  *  *  *  *  *
*  *  *  *  *  *  *  *  *  *  *  *  *  *  Q  *  *  *  *  *
*  *  *  *  *  *  *  *  Q  *  *  *  *  *  *  *  *  *  *  *
*  *  *  *  *  *  *  *  *  *  *  *  *  *  *  *  *  Q  *  *
*  Q  *  *  *  *  *  *  *  *  *  *  *  *  *  *  *  *  *  *
*  *  *  Q  *  *  *  *  *  *  *  *  *  *  *  *  *  *  *  *
*  *  *  *  *  *  *  *  *  *  *  *  *  *  *  *  *  *  Q  *
*  *  *  *  *  *  *  *  *  *  *  Q  *  *  *  *  *  *  *  *
*  *  *  *  *  *  *  *  *  *  *  *  *  *  *  *  *  *  *  Q
*  *  *  *  *  Q  *  *  *  *  *  *  *  *  *  *  *  *  *  *
*  *  *  *  *  *  *  *  *  *  *  *  *  Q  *  *  *  *  *  *
*  *  *  *  *  *  *  *  *  *  Q  *  *  *  *  *  *  *  *  *
*  *  *  *  *  *  *  Q  *  *  *  *  *  *  *  *  *  *  *  *
*  *  *  *  *  *  *  *  *  *  *  *  *  *  *  Q  *  *  *  *
*  *  Q  *  *  *  *  *  *  *  *  *  *  *  *  *  *  *  *  *
*  *  *  *  *  *  *  *  *  *  *  *  *  *  *  *  Q  *  *  *
*  *  *  *  *  *  Q  *  *  *  *  *  *  *  *  *  *  *  *  *
*  *  *  *  *  *  *  *  *  Q  *  *  *  *  *  *  *  *  *  *
*  *  *  *  *  *  *  *  *  *  *  *  Q  *  *  *  *  *  *  *
```

Fig. 2.11 A sample solution for the 20-queen problem

To achieve these goals, six sample optimization problems are presented and discussed. First, two versions of the shortest path problem, one with an introductory formulation and the other with a more advanced formulation are solved. Then, two versions of the warehouse location allocation problem are presented. The third problem was the traveling salesperson which was solved using Miller-Tucker-Zemlin constraint. Finally, the N-queen problem was presented. In all cases, first, the aim of solving the problem is explained, then the mathematical formulation is presented, and finally, the Pyomo code is brought. The codes clearly show that Pyomo offers the advantage of enabling users to express equations in a manner closely resembling their mathematical form.

By knowing the topics presented in this chapter, and learning the mathematical model of the various components of the electricity supply and distribution system in the next chapter, the reader can write the mathematical model of an electricity network, translate it into Python and solve it with the help of Pyomo.

References

1. Hart WE, Laird C, Watson JP, Woodruff DL (2012) Pyomo: Optimization Modeling in Python, vol 67
2. Pyomo (2023) Pyomo installation guide. Readthedocs. https://pyomo.readthedocs.io/en/stable/installation.html. Accessed 29 Dec 2023
3. Miller CE, Tucker AW, Zemlin RA (1960) Integer programming formulation of traveling salesman problems. J ACM 7(4):326–329. https://doi.org/10.1145/321043.321046

4. AMPL (2024) Free open-source solvers for AMPL. AMPL. https://ampl.com/products/solvers/open-source-solvers/. Accessed 15 Sep 2023

Chapter 3
Electricity System Modeling

Abstract Thermal power plants are the backbone of the power system supply side. They burn fuel, either fossil fuel or different types of renewable fuels such as biomass, and generate electricity which is then sent to the demand side using transmission and distribution lines. To optimize supply side performance, first the objective function should be defined clearly and then important operational constraints must be specified and modelled mathematically. The aim of this chapter is the introduce the objective function and main technical and operational constraints of thermal power plants and provide their mathematical formulation.

3.1 Introduction

Electricity or power system consists of three main parts; Firstly, there is the generation part in which electricity is generated by the power plants. The power plants could be thermal and use steam or gas turbines (or a combination of both), or they can generate power using renewable sources such as wind, hydro, and solar energy. The majority of thermal power plants burn fossil fuels for electricity generation; however, it is possible to use other sources of heat like solar energy (e.g., concentrated solar plants) or biofuels. The next step is to send the power from the plants to the consumers, it is done using the transmission and distribution lines. In a very basic formulation, one might neglect the existence of the transmission line and try to optimize the generation units considering that the demand should be satisfied; Nonetheless, this chapter considers the inefficiencies and costs associated with the transmission lines. Finally, the consumer or the demand side receives the electricity and converts it to useful energy (e.g., light, cooling, heating, etc.). The overall structure is shown in Fig. 3.1.

The schematic shown in Fig. 3.1 is very general and there are details that should be taken into account when modeling the electricity system. This chapter will not dive into the smallest level of details (such as frequency, active and reactive powers, voltage, and current equations) but tries to provide an overall understanding of

A. Ghadertootoonchi et al., *Energy System Modeling and Optimization*, SpringerBriefs in Energy, https://doi.org/10.1007/978-3-031-65906-5_3

Fig. 3.1 The overall structure of the power system

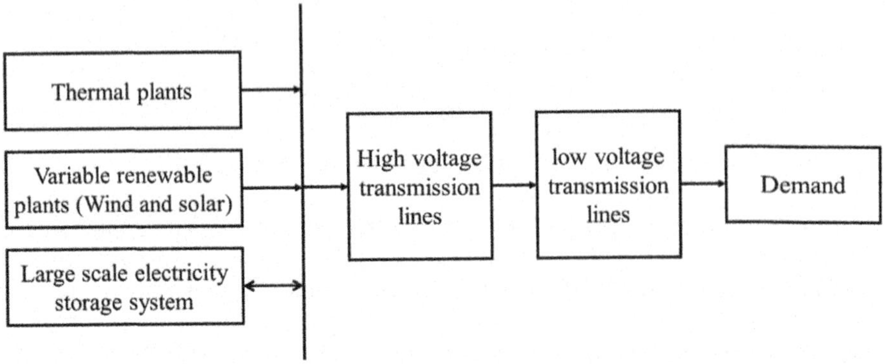

Fig. 3.2 A more detailed version of the overall structure of the power system

how electricity generation companies utilize optimization models to minimize their operational costs. A more accurate representation of Fig. 3.1 is brought in Fig. 3.2.

Figure 3.2 separates electricity generation technologies into thermal and variable renewable plants. Additionally, the transmission line is divided into high voltage and low voltage. Although some consumers require high-voltage electricity, here it is assumed that the demand is located at the low-voltage end of the schematic. However, the power plants are located in different geographical locations. Since the aim of the chapter is to provide a scalable and flexible formulation, we should be able to take into account these complexities. To do so, one can use the schematic depicted in Fig. 3.3.

Figure 3.3 represents a graph-based structure for the power system which is more decentralized than the previous two versions. In this structure, the generation units are spread on the map and they are connected by the transmission lines. The generation and consumption locations can be considered as the nodes of the graph while the transmission lines are the graph edges. A node can be a generation, consumption, or a prosumer, meaning that it can generate and consume electricity at the same time. Figure 3.4 shows different types of nodes in the modeling problem.

Nodes can also be with or without storage units. In any case, the node must follow Eq. (3.1) which is known as the energy conservation constraint.

$$I + G = O + L + D \tag{3.1}$$

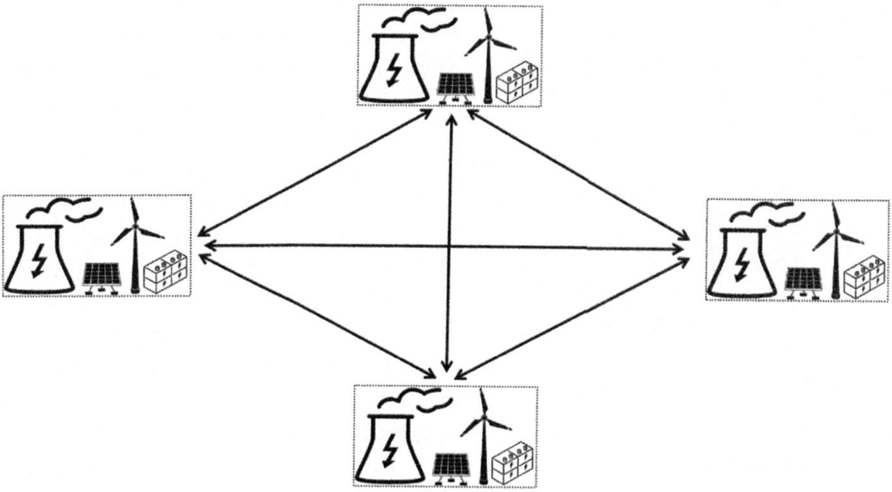

Fig. 3.3 Graph-based power system structure

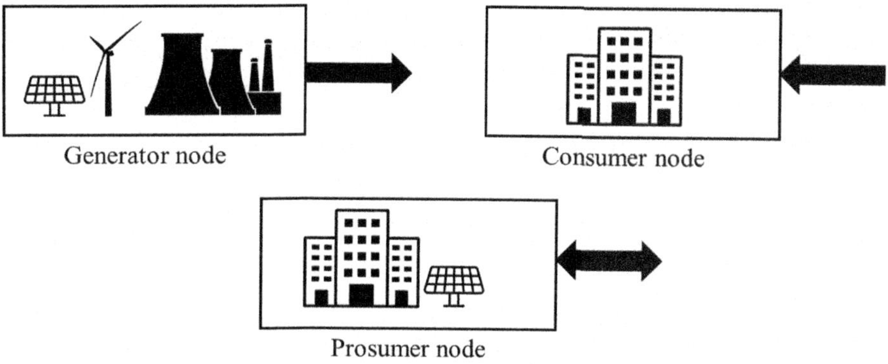

Fig. 3.4 Generator, consumer, and prosumer nodes

where I, is the energy input, G is the energy generation, O is the energy output, L is the energy loss due to inefficiencies, and D, is the energy demand. Additionally, although Fig. 3.3 indicates that all nodes feature thermal power plants, VRE (i.e., wind and solar), and EES, the actual composition of energy conversion units at each location may vary. Similarly, all the possible connection between the nodes is presented in Fig. 3.3, but only some of them may exist in a real system.

To optimize the considered power system, we should first know who is trying to optimize the system. One can formulate the problem from the demand side perspective, or considering the cost and income of power generation companies. In addition, transmission lines are usually controlled by independent transmission systems operators (TSO) and their interests must also be taken into account. Here, it is assumed that the objective is to minimize the cost of electricity generation; Hence, the model is developed considering the supply-side perspective. In addition, it is considered that the supply side should pay the TSO for every MWh of electricity it sends through the transmission lines. TSO transmits the generated electricity as long as the lines are not overloaded.

Another assumption is that all the generation units, no matter where their locations are, are controlled by a single company. In reality, multiple companies may compete to convince a consumer to buy their electricity. This case is discussed in this chapter but the focus is on the first assumption. Finally, the optimization problem discussed in this book is based on scheduling or optimization of the existing system operation. Planning or investing in new technologies, expanding the transmission lines, and changing the capacity of the generation units is not considered.

To start developing the mathematical formulation of the power system optimization we first need to know thermal power plants' characteristics, then we need to know what is the objective function and after defining the OF, we should pay attention to the constraints of the problem. These constraints are related to the supply side, transmission lines, and the demand side. At the end of this chapter, the reader will be able to understand the mathematical formulation of the power system optimization problem and will be able to follow Chap. 4, in which all the presented concepts are translated into Pyomo, and the optimization results are presented and discussed.

3.2 Thermal Power Plants

Thermal power plants burn fuel and produce hot gas or steam which then rotates the shaft of a gas or steam turbine. The rotation turns into electricity in a generator. Gas turbines usually follow the Brayton cycle shown in Fig. 3.5a, and steam turbines follow the Rankin cycle shown in Fig. 3.5b [1].

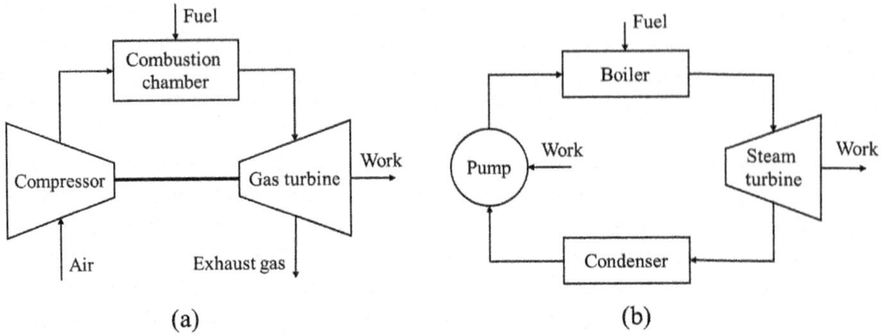

Fig. 3.5 **a** Brayton cycle for power generation using gas turbines and **b** Ranking cycle for power generation using steam turbines

Note that the term "gas turbine" does not mean that the fuel used in the Brayton cycle is natural gas, rather, it implies that in all the stages of the cycle, the working fluid is in the gaseous state. This is in contrast to the Ranking cycle in which there are water and steam at the same time.

While for some applications it is necessary to simulate the thermodynamic performance (e.g., energy and exergy efficiency) of these thermal units, it is out of the scope of this book. Here, we are curious about the operational cost of these units so that we can use them in the most cost-efficient way. The operational cost of these units is mainly a result of their fuel consumption. Therefore, one can use Eq. (3.2) to estimate the operational cost of a thermal unit.

$$OC = C_f \times FC \tag{3.2}$$

where OC is the operational cost, C_f is the cost of fuel ($\$/MW$), and FC is the fuel consumption (MW). However, this formulation only considers the fuel cost. Thermal power plants also have an operation and maintenance cost, this cost must be paid to ensure that the unit will work as expected and to maintain reliability. Equation (3.3) takes this cost into account.

$$OC = C_{fix} + C_{fuel} \times FC \tag{3.3}$$

where C_{fix} is the fixed maintenance cost. Moreover, we may not be aware of the amount of fuel being consumed by the thermal unit, but we know the amount of power generated by the unit. In this case, Eq. (3.4) can be used to estimate the operational cost.

$$OC = C_{fix} + C_f \times \frac{P}{\eta} \tag{3.4}$$

where P is the electrical power generated by the unit (MW) and η is the efficiency of the thermal unit ($FC = P/\eta$). Note that every thermal unit has an upper and lower bound for power generation. Therefore, the constraint shown in Eq. (3.5) must be taken into account.

$$P_{min} \leq P \leq P_{max} \tag{3.5}$$

However, it has been shown that the efficiency of thermal power plants is not constant and is a second-order function of load ratio ($actualload/nominalload$) [2, 3]. The relationship is shown in Fig. 3.6. This means that, the more the load ratio, the more the power generation, and the more the efficiency of the turbine ($\eta = f(P)$).

Considering the second-order relation, the final function that is used to calculate the OC is presented in Eq. (3.6) [4, 5].

$$OC = a \times P^2 + b \times P + c \tag{3.6}$$

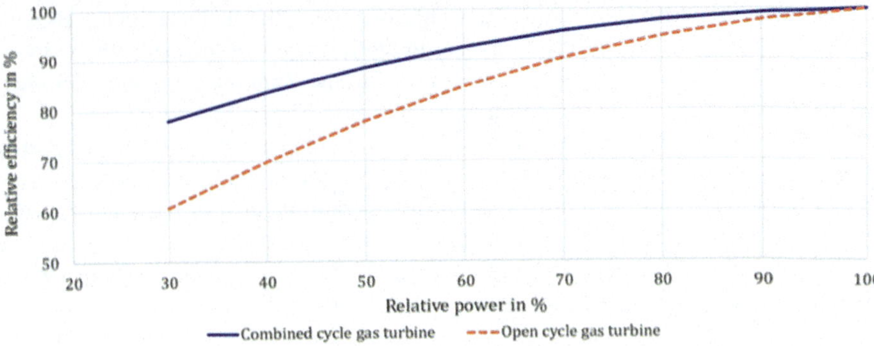

Fig. 3.6 Relative efficiency (current efficiency as a share of maximum efficiency) as a function of load ratio [2]

Although Eq. (3.6) is widely used to estimate the economic cost of operating a thermal unit, it does not represent the unit's GHG emission. To bridge this gap, we need an equation similar to Eq. (3.7) that estimates GHG emissions as a function of power generation. To do so, one can utilize Eq. (3.7) [6–8].

$$EM = \gamma \times P^2 + \beta \times P + \alpha + \mu \times \exp(\sigma \times P) \tag{3.7}$$

Finally, one needs to know the price of GHG emission ($\$/ton$) to convert the emission given by Eq. (3.7) to the monetary value and combine Eqs. (3.5) and (3.6) to get the total operational cost as shown in Eq. (3.8).

$$TOC = OC + C_{GHG} \times EM \tag{3.8}$$

Equation (3.8) provides enough information to calculate thermal units' operational costs. Nonetheless, there are other costs associated with thermal power generation units. These are discussed in the next section.

3.3 Objective Function

As mentioned in the previous section, the objective of the problem is to minimize the operational costs of the power generation units. However, GHG emissions and their associated costs must also be considered. Since the aim is to minimize economic and environmental costs, the optimization problem can be called an eco-environmental dispatch problem. The objective function of the problem is brought in Eq. (3.9).

$$\begin{aligned} \min OF = {} & \omega_c \times (SUSDC + OC + EEC + LLC + TC - PI - RI) \\ & + \omega_e \times (SUSDEC + OEC) \end{aligned} \tag{3.9}$$

where ω_c and ω_e represent the weight factors of the bi-objective optimization and $\omega_c + \omega_e = 1$. The first term represents the economic costs (with positive signs) and incomes (with negative signs). The second term represents the environmental costs of the power system. The definition of every term is brought in the following.

Thermal units need to be preheated before turning on and one cannot suddenly shut them down since they need to be cooled. During the preheating and cooling stages, fuel is burned without having power output. The process is both costly and emits GHG. The *SUSDC* is the start-up (SU) and shut down (SD) economic cost while *SUSDEC* is the start-up and shut down emission cost. In addition, when the unit is working, it consumes fuel. Similar to the previous case, one must pay for the fuel and the associated emissions. These are represented by the operational costs (*OC*) and the operational emission costs (*OEC*). The *EEC* is the cost of generating excess electricity. As mentioned in the previous section, TSO continues sending electricity from the supply side to the demand side as long as the lines are not overloading. If generators produce excess electricity, there would be a penalty that must be paid by the supply side. Similarly, if the generators produce less electricity than the demand and fail to satisfy the demand, there would be a penalty that they should pay represented by the load loss cost (*LLC*). In addition, TSO charges the supply side for providing transmission capacity which is the transmission cost (*TC*). Finally, there are two sources of financial income for the supply side. First one is the income from selling power (*PI*), and second one is the income from providing reserve capacity to maintain the reliability of the network (*RI*). *SUSDC* and *SUSDEC* formulations are brought in Eqs. (3.10) and (3.11).

$$SUSDC = \sum_{g \in G, t \in T, n \in N} SUC_g \times v_{g,t,n} + SDC_g \times w_{g,t,n} \tag{3.10}$$

$$SUSDEC = \sum_{g \in G, t \in T, n \in N} EMC_n \left(SUE_g \times v_{g,t,n} + SDE_g \times w_{g,t,n} \right), \tag{3.11}$$

where SUC_g and SDC_g are the SU and shutdown costs of generator of type g. The G is the set of generators (there could be different models of generators with different characteristics), T is the set of time steps (e.g., from hour 1 to 24), and N is the set of nodes (e.g., in Fig. 3.3 there are four nodes from 1 to 4). Equation (3.4) shows the SU and SD emission costs; EMC_n is the emission cost ($\$/ton$) in node n, SUE_g is the SU emission of generator g while SDE_g is the SD emission of generator g. $v_{g,t,n}$ and $w_{g,t,n}$ are binary decision variables that will be defined in the constraints subsection. The *OC* and *OEC* are computed based on Eqs. (3.12) and (3.13), as discussed in the previous section.

$$OC = \sum_{g \in G, t \in T, n \in N} a_g \times P_{g,t,n}^2 + b_g \times P_{g,t,n} + c_g \tag{3.12}$$

$$OEC = \sum_{g \in G, t \in T, n \in N} EMC_n \times \left(\gamma_g \times P_{g,t,n}^2 + \beta_g \times P_{g,t,n} + \alpha_g + \mu_g \times e^{\sigma_g \times P_{g,t,n}} \right), \tag{3.13}$$

where $P_{g,t,n}$ is a decision variable showing the power output of generator type g in node n at time t and it is limited by the maximum and minimum power output of the generator. In Eq. (3.12), a ($\$/MW^2$), b ($\$/MW$), and c (\$) are regression coefficients. These constants are different for different turbines and input fuels and can be determined by having data from the units' operation. Similarly, γ (ton/MW^2), β (ton/MW), α (ton), μ (ton), and σ (ton/MW) are regression constants obtained from the curve fitting. The exponential term is added because when the power output is near the maximum, combustion may occur incompletely and the amount of GHG emission experiences a higher growth rate. The *EEC* and *LLC* are calculated based on Eqs. (3.14) and (3.15).

$$EEC = \sum_{t \in T, n \in N} EC_n \times EE_{t,n} \tag{3.14}$$

$$LLC = \sum_{t \in T, n \in N} LC_n \times LL_{t,n}, \tag{3.15}$$

where EC_n and CC_n show the penalty for generating electricity more and less than demand in node n, respectively. The $EE_{t,n}$ is the excess electricity generation while $LL_{t,n}$ is the amount of lost load in node n at time t. The EC_n and LC_n are penalty parameters for excess generation and load loss ($\$/MW$). To calculate the transmission cost, one can use Eq. (3.16).

$$TC = \sum_{t \in T, n \in N, n' \in N, n \neq n'} TRC_{n,n'} \times f_{t,n,n'}, \tag{3.16}$$

where $TRC_{n,n'}$ is the cost of sending electricity from node n to node n' (parameter) and $f_{t,n,n'}$ is a decision variable showing the amount of electricity flow between these two nodes at time t such that $n \neq n'$. Finally, to calculate the supply side income, Eqs. (3.17) and (3.18) should be used.

$$PI = \sum_{t \in T, n \in N} PP_t \times \left(P_{t,n}^{th,con} + P_{t,n}^{ren,con} \right) + \sum_{t \in T, n \in N, s \in S} PP_t \times Dis_{s,t,n} \tag{3.17}$$

$$RI = \sum_{t \in T, n \in N} RP_t \times R_{g,t,n}, \tag{3.18}$$

where PP_t is the power price at time t (it could be different from one node to another, here the price is the same for all the nodes), $P_{t,n}^{th,con}$ and $P_{t,n}^{ren,con}$ are the amount of thermal and renewable power consumed in node n at time t. Note that due to the possibility of having excess generation, income must not be calculated based on $P_{g,t,n}$. The $Dis_{s,t,n}$ is the electricity discharged from storage type s. Moreover, RP_t is the reserve price at time t and $R_{g,t,n}$ is the reserve capacity provided by generator g.

All OF terms are introduced and discussed in this section, the relationship between the decision variables is explained in the constraints section, but before that, it is worth mentioning that the OF is a nonlinear function due to the presence of Eqs. (3.12) and (3.13). To formulate the model as a MILP, these equations must be linearized. The linearization technique is discussed in the following section.

3.4 Operational Cost Linearization

The piece-wise linearization method is employed to linearize the operational cost. In this method, the range between the minimum and maximum power output of the generator is divided into several subsections with a similar length which depends on the number of subsections. Then a line will be fitted on the nonlinear curve at each sub-section [9]. Figure 3.7 shows the piece-wise linear approximation of a nonlinear function.

In this method, if one considers $l_{max} + 1$ points for creating the subsections, l_{max} lines is fitted on the nonlinear curve. The slope of the lines can be calculated based on Eq. (3.19).

$$M_l = \frac{y_{l+1} - y_l}{P_{l+1} - P_l} \quad l \in \{1, \ldots, l_{\max}\} \tag{3.19}$$

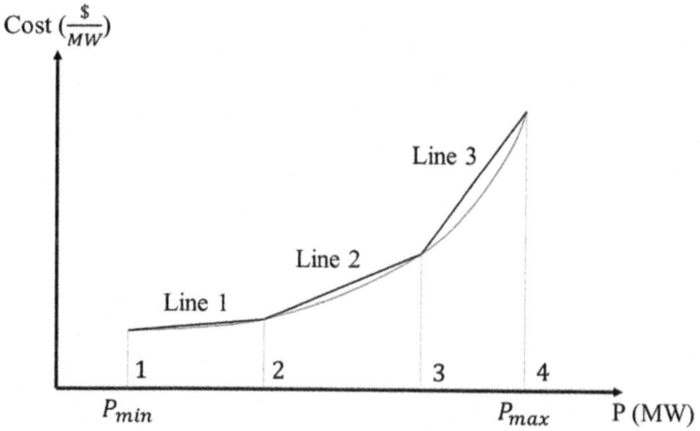

Fig. 3.7 The piece-wise linear approximation of a nonlinear generator cost function

where y_l is the value of the nonlinear function at point P_l. Additionally, the length of each subsection can be calculated based on Eq. (3.20).

$$Len = \frac{P_{max} - P_{min}}{l_{max}} \tag{3.20}$$

A set, called L, is used to represent the fitted lines defined as $L = \{1,2,3\}$. Therefore, l_1 is the first fitted line, l_2 is the second fitted line, and so on. Note that Len has different values for different generators since P_{min} and P_{max} are different. To linearize the nonlinear function, Eqs. (3.21) to (3.24) can be used.

$$P_{g,t,n} = P_{\min,g} \times u_{g,t,n} + \sum_{l \in L} P_{l,g,t,n} \quad \forall g \in G, t \in T, n \in N \tag{3.21}$$

$$P_{l,g,t,n} \leq Len_g \times u_{g,t,n} \quad \forall g \in G, t \in T, n \in N, l \in L \tag{3.22}$$

$$OC = \sum_{g \in G, t \in T, n \in N} \left(OC_{\min,g} \times u_{g,t,n} + \sum_{l \in L} \left(M_{g,l}^{cost} \times P_{l,g,t,n} \right) \right) \tag{3.23}$$

$$OEC = \sum_{g \in G, t \in T, n \in N} EMC_n \times \left(EM_{\min,g} \times u_{g,t,n} + \sum_{l \in L} \left(M_{g,l}^{emsn} \times P_{l,g,t,n} \right) \right) \tag{3.24}$$

Equation (3.21) states that the power output of a generator is the sum of the linear sub-section powers. The power generation at each linear subsection must not exceed the length of the sub-section as shown in Eq. (3.22). The $u_{g,t,n}$ is a binary variable that determines whether the generator is working (1) or not (0). Equation (3.23) is the linearized version of Eq. (3.12). If the generator is working ($u_{g,t,n} = 1$), a constant cost ($OC_{min,g}$) must be paid which is associated to $P_{min,g}$. If the power output increases, the cost will increase considering the slope of the fitted line ($M_{g,l}^{cost}$). Similarly, if the generator is working, there will be a minimum amount of emission equal to $EM_{min,g}$, and when the generation is more than $P_{min,g}$, the emission increases with a slope of $M_{g,l}^{emsn}$. Figure 3.8 is useful to clarify these explanations.

It can be seen that C_L overestimates the cost of generating $P(MW)$ electricity. While overestimation is a drawback of the utilized linearization method, it can be reduced by increasing the number of fitted lines which increases the number of decision variables. Now, we can use the linearized function in the optimization problem, and in the next sections, the constraints of the problem are discussed.

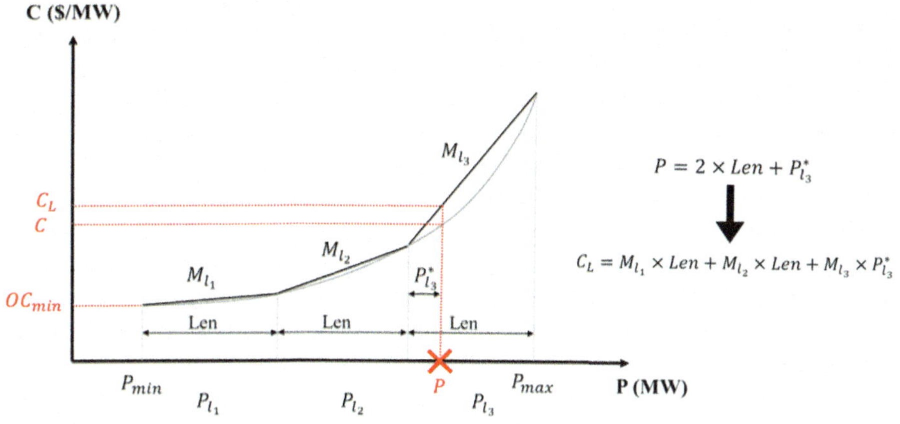

Fig. 3.8 The process of calculating the linearized cost

3.5 Optimization Model Constraints

This section will describe the mathematical formulation of the MILP optimization model constraints. The constraints are categorized into 12 different groups. It is worth mentioning that all the decision variables of the model are nonnegative real numbers except for $v_{g,t,n}$, $w_{g,t,n}$, $u_{g,t,n}$, $\chi_{t,n,nn}$, $B^{Ch}_{s,t,n}$, and $B^{Dis}_{s,t,n}$ which are binary variables. All the decision variables are presented and discussed in the next section.

3.5.1 Start-Up and Shut-Down

The start-up constraints aim to establish a logical connection between $u_{g,t,n}$ and $v_{g,t,n}$, whereas the shut-down constraint introduce the connection between $u_{g,t,n}$ and $w_{g,t,n}$. The initial value of these variables is zero and they can be calculated based on Eqs. (3.25) and (3.26).

$$u_{g,t,n} - u_{g,t-1,n} \leq v_{g,t,n} \quad \forall g \in G, t \in T - \{t_{start}\}, n \in N, \tag{3.25}$$

$$u_{g,t-1,n} - u_{g,t,n} \leq w_{g,t,n} \quad \forall g \in G, t \in T - \{t_{start}\}, n \in N. \tag{3.26}$$

These equations force the model to start a generator if $u_{g,t-1,n} = 0$ AND $u_{g,t,n} = 1$, and to shut it down if $u_{g,t,n} = 1$ AND $u_{g,t-1,n} = 0$. For $t = t_{start}$, the start-up constraint is $u_{g,t,n} \leq v_{g,t,n}$, and since the initial value of the binary variables is zero, there is no need to define the shut-down constraint for $t = t_{start}$. Table 3.1 shows possible values of these three binary variables.

Table 3.1 Possible values for $u_{g,t,n}$, $v_{g,t,n}$, and $w_{g,t,n}$

State	$u_{g,t-1,n}$	$u_{g,t,n}$	$u_{g,t,n} - u_{g,t-1,n}$	$v_{g,t,n}$	$w_{g,t,n}$
Steady state (on)	1	1	0	0	0
Turn off	1	0	−1	0	1
Turn on	0	1	1	1	0
Steady state (off)	0	0	0	0	0

Due to the fact that there are SU and SD costs and emissions, the preferred value for $v_{g,t,n}$ and $w_{g,t,n}$ is zero. This means the model does not turn on or off generators unless it is necessary. Another important constraint is the minimum up and down time described in the following.

3.5.2 *Minimum up and Down Time*

If the model decides to start generator g in node n at time t, then it should be on for at least minimum up time; The mathematical representation of such a constraint is brought in Eq. (3.27).

$$\sum_{t^* \in T^*} u_{g,t^*,n} \geq \left|T^*\right| \times v_{g,t,n} \quad \forall g \in G, n \in N,$$
$$t \in T, t^* \in T^*,$$
$$T^* = \left\{t, \ldots, \min\left(|T|, t + MUT_g - 1\right)\right\}. \tag{3.27}$$

Equation (3.27) states that if the generator is turned on at time t, it should be on during T^* (a subset of T). T^* upper bound is a function of the optimization horizon ($|T|$ or number of elements in T) and minimum operation time (MUT_g). The -1 is included since time t itself should be included in the operating hours. Similarly, if the model decides to turn the generator off, it should be off for at least its minimum downtime (MDT_g). The related constraint is shown in Eq. (3.28).

$$\sum_{t^* \in T^*} 1 - u_{g,t^*,n} \geq \left|T^*\right| \times w_{g,t,n} \quad \forall g \in G,$$
$$n \in N, t \in T, t^* \in T^*,$$
$$T^* = \left\{t, \ldots, \min\left(|T|, t + MDT_g - 1\right)\right\} \tag{3.28}$$

Therefore, if $w_{g,t,n} = 1$, then $u_{g,t^*,n}$ must be zero during T^*. Each generator has a minimum and maximum power output which is introduced in the next section.

3.5.3 Thermal Power Generation Bound

As mentioned in Sect. 3.3, every generator has a maximum and minimum power output and $P_{g,t,n} \in [P_{min,g}, P_{max,g}]$. However, due to the presence of Eq. (3.21), there is no need to add this constraint to the problem since Eq. (3.21) ensures that $P_{g,t,n}$ is between $P_{min,g}$ and $P_{max,g}$ or is zero if the system is off ($U_{g,t,n} = 0$).

3.5.4 Ramp up and Down

If a generator is off ($P_{g,t,n} = 0$), and it is turned on at time t ($v_{g,t,n} = 1$), then the output will be equal to P_g^{min}. After that, the increase in output is limited by the generators ramp up rate. Such a condition can be modeled by Eq. (3.29).

$$\begin{aligned} P_{g,t,n} - P_{g,t-1,n} \leq RU_g\left(1 + u_{g,t-1,n} - u_{g,t,n}\right) \\ + P_g^{\min}\left(2 - u_{g,t,n} - u_{g,t-1,n}\right) \quad \forall g \in G, \\ n \in N, t \in T - \{t_{start}\} \end{aligned} \tag{3.29}$$

For $t = t_{start}$, the constraint is $P_{g,t,n} = u_{g,t,n} \times P_g^{min}$. The RU_g (MW/h) is the ramp up rate of generator g. Table 3.2 shows different modes of Eq. (3.29).

Similarly, if a generator is on at time $t - 1$, and it is turned off at time t, then the generator output at time $t - 1$ must be equal to P_g^{min}. In working condition, a decrease in the generator's output must be less than the ramp down rate as shown in Eq. (3.30).

$$\begin{aligned} P_{g,t-1,n} - P_{g,t,n} \leq RD_g\left(1 - u_{g,t-1,n} + u_{g,t,n}\right) \\ + P_g^{\min}\left(2 - u_{g,t,n} - u_{g,t-1,n}\right) \\ \forall g \in G, n \in N, t \in T - \{t_{start}\}, \end{aligned} \tag{3.30}$$

where RD_g is the ramp down rate of generator g (MW/h). In the beginning, all generators are off, so there is no need for this constraint. Table 3.3 shows different modes of the ramp down rate constraint.

Table 3.2 Different modes of the ramp up rate constraint

$u_{g,t-1,n}$	$u_{g,t,n}$	Right-hand side
0	0	Trivial constraint
0	1	P_g^{min} (i.e., upon start $p_{g,t,n} = P_g^{min}$)
1	0	Trivial constraint
1	1	RU_g

Table 3.3 Different modes of the ramp down rate constraint

$u_{g,t-1,n}$	$u_{g,t,n}$	Right hand side
0	0	Trivial constraint
0	1	P_g^{min} (i.e., upon start $p_{g,t,n} = P_g^{min}$)
1	0	P_g^{min} (upon shut-down $p_{g,t-1,n} = P_g^{min}$)
1	1	RD_g

3.5.5 Electricity Balance

Electricity balance constraint, in a simple form, states that the generation must be exactly equal to the demand as shown in Eq. (3.31).

$$P_{t,n}^{th} + P_{t,n}^{ren} = D_{t,n} \quad \forall t \in T, n \in N, \tag{3.31}$$

where $P_{t,n}^{th}$ is the power generated by the thermal units, $P_{t,n}^{ren}$ is the power generated by the renewable units, and $D_{t,n}$ is the electricity demand. However, when the aim is to model a network in which there is a trade of energy between the nodes, import and export must also be considered in the balance constraint as shown in Eq. (3.32).

$$P_{t,n}^{th,con} + P_{t,n}^{ren,con} + Trn_{t,n} = D_{t,n} \quad \forall t \in T, n \in N, \tag{3.32}$$

where $Trn_{t,n}$ is the net effect of electricity trade (positive: import, negative: export). In addition, there could be periods during which the supply side cannot satisfy the exact amount of electricity demand. Therefore, one should consider the effect of load loss and excess electricity generation. Nevertheless, to maintain the network stability, load loss and excess electricity generation must be minimized which is the reason behind the addition of these two variables to the OF. Finally, the effect of energy storage systems should be considered in the constraint. Balance constraint's final version is brought in Eq. (3.33).

$$P_{t,n}^{th,con} + P_{t,n}^{ren,con} + St_{t,n} + Trn_{t,n} = D_{t,n} - LL_{t,n} \quad \forall t \in T, n \in N. \tag{3.33}$$

Equation (3.33) states that the summation of thermal [Eq. (3.34)] and renewable [Eq. (3.35)] electricity consumed in node n, the net effect of storage units [Eq. (3.36)], and net import [Eq. (3.37)] must be equal to the demand minus the load loss ($LL_{t,n}$). Renewable energy generation can be imported to the model from external sources, the process will be described in the next chapter.

$$\sum_{g \in G} P_{g,t,n} = P_{t,n}^{th,con} + P_{t,n}^{th,exs} \quad \forall t \in T, n \in N \tag{3.34}$$

$$\sum_{r \in R} P_{r,t,n} = P_{t,n}^{ren,con} + P_{t,n}^{ren,exs} \quad \forall t \in T, n \in N \tag{3.35}$$

$$St_{t,n} = \sum_{s \in S} \left(Dis_{s,t,n} - Ch_{s,t,n}\right) \quad \forall t \in T, n \in N \tag{3.36}$$

$$Trn_{t,n} = \sum_{n' \in N} \left(f_{t,n',n} \times \eta_{n',n} - f_{t,n,n'}\right) \quad \forall t \in T, n \in N, n \neq n' \tag{3.37}$$

where Eqs. (3.33) and (3.34) divide the thermal and renewable electricity into two parts of excess and consumed, respectively. The R is the set of renewable energy technologies. Equation (3.36) indicates the net effect of the storage units (positive for discharging and negative for charging). Equation (3.37) represents the same concept for the transmitted electricity considering the transmission lines' efficiency.

It should be noted that if generator type g does not exist in node n, all the related value to the generator g in node n is fixed to zero. The same is true for renewable type r and storage type s in node n. Also, as mentioned in Sect. 3.3, a node cannot send electricity to itself and $f_{t,n,n} = 0.$. Finally, the total amount of excess electricity which should be minimized [see Eqs. (3.14) and (3.9)] is calculated based on Eq. (3.38).

$$EE_{t,n} = P_{t,n}^{th,exs} + P_{t,n}^{ren,exs} \tag{3.38}$$

In the next section, the constraints imposed on the excess electricity and the load loss are demonstrated.

3.5.6 Excess Electricity and Load Loss

Six constraints are considered for the upper bounds of the excess and curtailed electricity which will be described in this section. There is an upper bound for the hourly excess electricity and load loss shown in Eqs. (3.39) and (3.40). The upper bound is regional since the power stability of one region could be more important than the other.

$$EE_{t,n} \leq EE_{n,hourly}^{\max} \quad \forall n \in N, t \in T \tag{3.39}$$

$$LL_{t,n} \leq LL_{n,hourly}^{\max} \quad \forall n \in N, t \in T. \tag{3.40}$$

In addition to the hourly upper bound, there is an upper bound on the total amount of excess and curtailed electricity in every region represented by Eqs. (3.41) and (3.42).

$$\sum_{t \in T} EE_{t,n} \leq EE_n^{\max} \quad \forall n \in N \tag{3.41}$$

$$\sum_{t \in T} LL_{t,n} \leq LL_n^{\max} \quad \forall n \in N \tag{3.42}$$

Finally, the upper bound for total excess (EE^{max}) and curtail electricity (LL^{max}) over the entire network is shown in Eqs. (3.43) and (3.33).

$$\sum_{t\in T,n\in N} EE_{t,n} \leq EE^{\max} \tag{3.43}$$

$$\sum_{t\in T,n\in N} LL_{t,n} \leq LL^{\max} \tag{3.44}$$

The right-hand side of these equations is the optimization parameters. The next section will describe the electricity demand and demand response formulation.

3.5.7 *Demand Response*

To solve the model and get the optimal results, we need to know the electricity load the generators have to satisfy. The load can be obtained from external sources and imported to the optimization problem or it can be integrated into the model. Here, it is assumed that the electricity load is available from external sources. However, instead of directly using the electricity load, it is converted to electricity demand which is the actual power required by a region. It can be seen that $D_{t,n}$ is used in Eq. (3.33) rather than $L_{t,n}$ which is node n hourly load. The relationship between $L_{t,n}$ and $D_{t,n}$ is shown in Eqs. (3.45) and (3.46).

$$\sum_{t\in T} D_{t,n} = \sum_{t\in T} L_{t,n} \quad \forall n \in N \tag{3.45}$$

$$L_{t,n} \times (1 - \psi_{down}) \leq D_{t,n} \leq L_{t,n} \times \left(1 + \psi_{up}\right) \forall n \in N, t \in T \tag{3.46}$$

While $L_{t,n}$ is static, $D_{t,n}$ can change based on the generation costs. During peak hours, the model tries to reduce demand and during off-peak hours, it tries to increase the demand since the total value of demand and load are the same (Eq. (3.43)). This strategy is called the demand response (DR) and it increases the demand side flexibility. Equation (3.44) shows that $D_{t,n}$ can be increased by ψ_{up} or decreased by ψ_{down} both of which are values between 0 and 1.

3.5.8 *Transmission Lines*

Two constraints are imposed on transmission lines, and limit the maximum electricity flow between two nodes to prevent overloading. The formulation is brought in Eqs. (3.47) and (3.48).

$$f_{t,n,n'} \leq f_{n,n'}^{\max} \times \chi_{t,n,n'} \quad \forall t \in T, n \ \& \ n' \in N, \chi_{t,n,n'} \in \{0, 1\} \tag{3.47}$$

$$f_{t,n',n} \leq f_{n,n'}^{\max} \times \left(1 - \chi_{t,n,n'}\right) \quad \forall t \in T, n \,\&\, n' \in N, \chi_{t,n,n'} \in \{0, 1\}, \tag{3.48}$$

where $f_{n,n'}^{\max}$ is the transmission upper bound and $\chi_{t,n,n'}$ is a binary variable determining the direction of the energy flow. If $\chi_{t,n,n'} = 1$, electricity is flowing from n to n', otherwise, the direction of flow is from n' to n. These equations assume that the transmission lines' reliability is 100% and there is no possibility for transmission line outages. To account for the effects of scheduled maintenance or any event causing a power line outage on optimal performance, you can introduce a binary parameter, denoted as $Out_{t,n,n'}$, where it equals 1, if the line is operational and 0 otherwise. It is possible to optimize this new parameter to identify the optimal maintenance period that has the least impact on the network's operation. Final transmission equations are brought in Eqs. (3.49) and (3.50).

$$f_{t,n,n'} \leq f_{n,n'}^{\max} \times Out_{t,n,n'} \times \chi_{t,n,n'} \quad \forall t \in T, n \ \& \ n' \in N, \chi_{t,n,n'} \in \{0, 1\} \tag{3.49}$$

$$f_{t,n',n} \leq f_{n,n'}^{\max} \times Out_{t,n,n'} \times \left(1 - \chi_{t,n,n'}\right) \quad \forall t \in T, n \ \& \ n' \in N, \chi_{t,n,n'} \in \{0, 1\} \tag{3.50}$$

The remaining transmission constraints are related to energy self-sufficiency described in the next section.

3.5.9 Energy Self-Sufficiency

If the aim is to make the nodes energy self-sufficient or neutralize their impact on the network over a period, the net-zero energy constraint can be added to the problem. The initial formulation of the constraint is presented in Eq. (3.51).

$$\sum_{n' \in N, t \in T} f_{t,n,n'} = \sum_{n' \in N, t \in T} \eta_{n',n} \times f_{t,n',n} \quad \forall n \ \& \ n' \in N, \tag{3.51}$$

where the left-hand side indicates the export and the right-hand side shows the import. The $\eta_{n',n}$ is the transmission line efficiency. The constraint states that the export from node n should be equal to the net electricity received by node n. To increase the flexibility of the constraint one can change it to an inequality, as shown in Eq. (3.52).

$$\sum_{n' \in N, t \in T} f_{t,n,n'} \geq \sum_{n' \in N, t \in T} \eta_{n',n} \times f_{t,n',n} \quad \forall n \ \& \ n' \in N. \tag{3.52}$$

Equation (3.52) represents a positive zero energy node where the export should be greater than or equal to the import. Finally, one can add a net-zero energy coefficient to the constraint and develop the final energy self-sufficiency constraint as brought in Eq. (3.53).

$$\sum_{n' \in N, t \in T} f_{t,n,n'} \geq NZEC_n \times \sum_{n' \in N, t \in T} \eta_{n',n} \times f_{t,n',n} \quad \forall n \;\&\; n' \in N. \tag{3.53}$$

The net-zero energy coefficient ($NZEC_n$) shows the percentage of being net-zero. For instance, if $NZEC_n = 0.2$, it means the electricity export of the node should exceed 20% of its import. When the coefficient is 1, the node is considered to be positive or net-zero energy and in other cases (except 0) it can be considered partially or nearly net-zero.

3.5.10 Electric Energy Storage System

As mentioned in Chap. 1 and Eq. (3.33), EES systems are added to the supply side to control the fluctuations of VREs and increase the network reliability. The role of EES units is becoming increasingly vital in today's electricity system as there is an emphasis on the effective utilization of VREs. EES systems can be modeled using different sets of equations. The basic version is brought in Eqs. (3.54)–(3.55).

$$\begin{aligned} SOE_{s,t,n} &= SOE_{s,t-1,n} + Ch_{s,t,n} \\ &\quad - Dis_{s,t,n} \forall s \in S, t \in T - \{t_{start}\}, n \in N \end{aligned} \tag{3.54}$$

$$SOE_{s,t,n} \leq Cap_s \quad \forall s \in S, \quad t \in T, \quad n \in N, \tag{3.55}$$

where $SOE_{s,t,n}$ is the state of energy or the amount of energy stored in the battery, $Ch_{s,t,n}$ is the energy input by charging, $Dis_{s,t,n}$ is the energy output by discharging, and Cap_s is the capacity of storage technology type s. While these equations represent the general operation of a storage unit, other limitations should be taken into account. First, charging and discharging energy should have an upper bound. Second, there should be a lower bound for $SOE_{s,t,n}$ to improve the battery performance and useful life. Third, the self-discharge effect must be considered. Finally, the storage cannot be charged or discharged simultaneously. These limitations are added to the constraints and brought in Eqs. (3.56)–(3.60) [10].

$$\begin{aligned} SOE_{s,t,n} &= \left(1 - \kappa_s^1\right) \times SOE_{s,t-1,n} + \eta_s^{ch} \\ &\quad \times Ch_{s,t,n} - \frac{1}{\eta_s^{Dis}} Dis_{s,t,n} - \kappa_s^2 \\ &\quad \forall s \in S, t \in T - \{t_{start}\}, n \in N \end{aligned} \tag{3.56}$$

$$\left(\frac{1-DOD_s^{\max}}{2}\right)\times Cap_s \le SOE_{s,t,n} \le \left(1-\frac{1-DOD_s^{\max}}{2}\right)\times Cap_s \quad \forall s\in S, t\in T,\ n\in N \tag{3.57}$$

$$Ch_{s,t,n}\le Ch_s^{\max}\times B_{s,t,n}^{Ch} \quad \forall s\in S, t\in T, n\in N, B_{s,t,n}^{Ch}\in\{0,1\} \tag{3.58}$$

$$Dis_{s,t,n}\le Dis_s^{\max}\times B_{s,t,n}^{Dis} \quad \forall s\in S, t\in T, n\in N, B_{s,t,n}^{Dis}\in\{0,1\} \tag{3.59}$$

$$B_{s,t,n}^{Ch}+B_{s,t,n}^{Dis}\le 1 \quad \forall s\in S, t\in T, n\in N, \tag{3.60}$$

where Eq. (3.56) shows the inter-time dependence of the stored energy [11, 12]. Equation (3.57) shows the acceptable SOE range based on the maximum allowable depth of discharge (DOD_s^{max}) and Cap_s. Equations (3.58) and (3.59) show the upper bound on charging and discharging. Two binary variables are used to prevent simultaneous charge and discharge based on Eq. (3.60). Note that the initial SOE (i.e., at t_{start}) should be introduced to the model. While EES units help increase the reliability of the network and facilitate effective integration of VREs, there is a need for another mechanism to ensure the supply side can continue providing power even if a thermal unit fails due to unprecedented reasons. This concept is called $N-1$ contingency and is discussed in the next section.

3.5.11 Reliability

To increase the network reliability, N − 1 contingency is considered. It ensures that even if one of the thermal generating units shows faulty behavior, other N − 1 generators can provide enough power for compensation [13, 14]. The constraints are brought in Eqs. (3.61)–(3.63).

$$r_{g,t,n}\le RU_g\times u_{g,t,n}+P_{min,g}\times\left(1-u_{g,t,n}\right)\forall t\in T, n\in N, g\in G \tag{3.61}$$

$$r_{g,t,n}\ge P_{min,g}\times\left(1-u_{g,t,n}\right) \tag{3.62}$$

$$r_{g,t,n}\le P_{max,g}-P_{g,t,n} \tag{3.63}$$

Table 3.4 clarifies the impact of these equations on the optimization problem. Note that considering Eqs. (3.21) and (3.22), if $u_{g,t,n}=0$, then $P_{g,t,n}=0$.

Table 3.4 Description of Eqs. (3.61)–(3.63)

$u_{g,t,n}$	Right hand side
1	$r_{g,t,n} \leq RU_g$ $r_{g,t,n} \geq 0$ $r_{g,t,n} \leq P_{\max,g} - P_{g,t,n}$ $=> 0 \leq r_{g,t,n} \leq \min(RU_g, P_{\max,g} - P_{g,t,n})$
0	$r_{g,t,n} \leq P_{\min,g}$ $r_{g,t,n} \geq P_{\min,g}$ $r_{g,t,n} \leq P_{\max,g}$ $=> r_{g,t,n} = P_{\min,g}$

As Table 3.4 shows, if the unit is working and $u_{g,t,n} = 1$, it can provide a maximum reserve equal to $\min(RU_g, P_{max,g} - P_{g,t,n})$, otherwise, the provided reserve is $P_{min,g}$. The problem with this formulation is that it forces the off units to provide a reserve capacity of $P_{min,g}$ and always be ready to start immediately. This could not always be the case. The off units can provide a reserve equal to $P_{min,g}$ or 0. To add this condition to the above constraints, another binary variable, $a_{g,t,n}$ is needed, as shown in Eqs. (3.64)–(3.67).

$$r_{g,t,n} \leq RU_g \times u_{g,t,n} + P_{min,g} \times \left(1 - u_{g,t,n} - a_{g,t,n}\right) \forall t \in T, n \in N, g \in G \quad (3.64)$$

$$r_{g,t,n} \geq P_{min,g} \times \left(1 - u_{g,t,n} - a_{g,t,n}\right) \forall t \in T, n \in N, g \in G \quad (3.65)$$

$$a_{g,t,n} \leq 1 - u_{g,t,n} \forall t \in T, n \in N, g \in G \quad (3.66)$$

$$r_{g,t,n} \leq P_{max,g} - P_{g,t,n} \forall t \in T, n \in N, g \in G. \quad (3.67)$$

If the unit is off and $a_{g,t,n} = 1$, then the reserve capacity is 0, otherwise, it is equal to 1. If the unit is on, then $a_{g,t,n}$ is forced to be 0, as shown in Eq. (3.66). Table 3.5 clarifies the impact of these equations on the optimization problem.

To calculate the reserve lower and upper bounds, Eqs. (3.68) and (3.69) are used.

$$\sum_{g' \in G, n' \in N} r_{g',t,n'} \geq P_{g,t,n} \quad \forall g \in G, n \in N, t \in T, g' \neq g \text{ OR } n' \neq n \quad (3.68)$$

$$\sum_{g' \in G, n' \in N} r_{g',t,n'} \leq P_{g,t,n} \times (1 + r_{max}) \quad \forall g \in G, n \in N, t \in T, g' \neq g \text{ OR } n' \neq n \quad (3.69)$$

where the left-hand side is the total available reserve capacity, and the right-hand sides represent the lower and upper bounds of the reserve capacity. The r_{max} is the percentage of reserve surplus to demand and has a value between 0 and 1. For

Table 3.5 Description of Eqs. (3.64)–(3.67)

$u_{g,t,n}$	$a_{g,t,n}$	Right hand side
1	0	$r_{g,t,n} \le RU_g$ $r_{g,t,n} \ge 0$ $r_{g,t,n} \le P_{\max,g} - P_{g,t,n}$ $=> 0 \le r_{g,t,n} \le \min(RU_g, P_{\max,g} - P_{g,t,n})$
0	1	$r_{g,t,n} \le 0$ $r_{g,t,n} \ge 0$ $r_{g,t,n} \le P_{\max,g}$ $=> r_{g,t,n} = 0$
	0	$r_{g,t,n} \le P_{\min,g}$ $r_{g,t,n} \ge P_{\min,g}$ $r_{g,t,n} \le P_{\max,g}$ $=> r_{g,t,n} = P_{\min,g}$

instance, if $r_{max} = 0.3$, the reserve capacity upper bound is 30% more than current power generation. Note that the summation in the above equations is over g' and n' which means summing over all the power generation of all generators except that of g. For clarification, assume there are three nodes each of which has a combination of thermal power units, as shown in Fig. 3.9.

If unit g_3 in node 3 fails (the red generator), then the sum of the reserve capacity of all other generators (the black ones) must be greater than or equal to the red generator power output. To find this value, we first sum over all generators in other nodes ($n' \neq n$), then we sum over all generators in n such that $g' \neq g$. Note that if

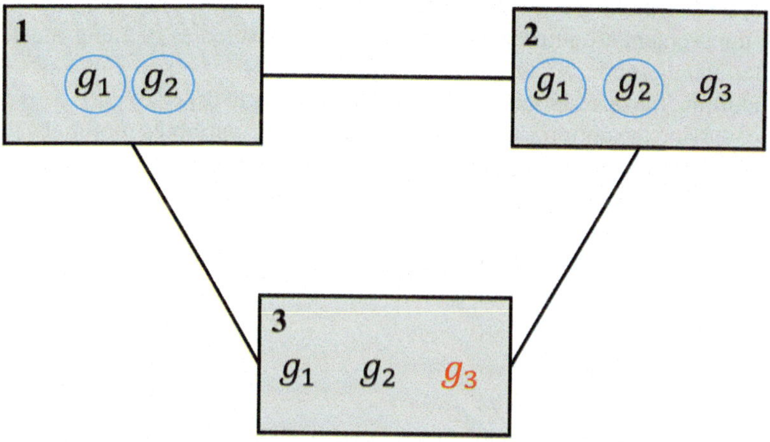

Fig. 3.9 Reserve capacity (sum of black generators' reserve) in case of generator failure (the red generator)

we use AND rather than OR, the summation will be over the generators with blue circles instead of the black ones. The last set of constraints belongs to the emission limits discussed in the next subsection.

3.5.12 *Emissions*

The last set of constraints of the model is the emission limits that consist of two constraints as shown in Eqs. (3.70) and (3.71).

$$\sum_{g\in G,t\in T,n\in N} \left(EM_{min,g} \times u_{g,t,n} + \sum_{l\in L} M_{g,l}^{emsn} \times P_{l,g,t,n} \right) \leq Em^{max} \tag{3.70}$$

$$\sum_{g\in G,t\in T} \left(EM_{min,g} \times u_{g,t,n} + \sum_{l\in L} M_{g,l}^{emsn} \times P_{l,g,t,n} \right) \leq Em_n^{max} \quad \forall n \in N, \tag{3.71}$$

where Eq. (3.70) shows the network's emission upper bound, whereas Eq. (3.71) shows the nodal emission upper bound. The next section presents the definition of decision variables.

3.6 Optimization Model Decision Variables

In the previous two sections, we discussed the OF and the constraints of the eco-environmental optimization problem used to minimize the operational costs of a combination of thermal and renewable power plants. Additionally, all the parameters of the problem are introduced. This section is devoted to defining the decision variables and their bounds as shown in Table 3.6.

In addition, the description of sets is brought in Table 3.7.

After solving the optimization problem, the decision variables' optimal value will be available for further investigation and sensitivity analysis. It is done in the next chapter, in which the case study is developed, and the parameters of the problem are introduced.

Table 3.6 Decision variable of the optimization problem and their explanation

Number	Decision variable	Domain	Bounds	Description
Objective function				
1	$SUSDC$	NNR	$[0, +\infty)$	Startup and shutdown costs
2	$SUSDEC$	NNR	$[0, +\infty)$	Startup and shut down emission costs
3	OC	NNR	$[0, +\infty)$	Operational costs
4	OEC	NNR	$[0, +\infty)$	Operational emission costs
5	EEC	NNR	$[0, +\infty)$	Excess electricity generation penalty
6	LLC	NNR	$[0, +\infty)$	Load loss penalty
7	TC	NNR	$[0, +\infty)$	Transmission cost
8	PI	NNR	$[0, +\infty)$	Income from selling power
9	RI	NNR	$[0, +\infty)$	Income from providing reserve capacity
Power generation and reserve capacity				
10	$P_{g,t,l,n}$	NNR	$[0, +\infty)$	Power generation in a linear section (fitted on the nonlinear cost function)
11	$P_{g,t,n}$	NNR	$[0, +\infty)$	Power generation
12	$r_{g,t,n}$	NNR	$[0, +\infty)$	Reserve capacity
Power flow between nodes				
13	$f_{n,n',t}$	NNR	$[0, +\infty)$	Power flow between two nodes
14	$\chi_{t,n,n'}$	Binary	$\{0,1\}$	Direction of power flow
Excess electricity and load loss				
15	$P_{t,n}^{ren,con}$	NNR	$[0, +\infty)$	Renewable power consumed in node n
16	$P_{t,n}^{ren,exs}$	NNR	$[0, +\infty)$	Excess renewable power in node n
17	$P_{t,n}^{th,con}$	NNR	$[0, +\infty)$	Thermal power consumed in node n

(continued)

Table 3.6 (continued)

Number	Decision variable	Domain	Bounds	Description
18	$P_{t,n}^{th,exs}$	NNR	$[0, +\infty)$	Excess thermal power in node n
19	$EE_{t,n}$	NNR	$[0, +\infty)$	Excess electricity at time t
20	$LL_{t,n}$	NNR	$[0, +\infty)$	Load loss at time t
Demand response				
21	$D_{t,n}$	NNR	$[L_{t,n}(1-\psi_{down}), L_{t,n}(1+\psi_{up})]$	Flexible demand is limited by the actual load
EES system				
22	$SOE_{s,t,n}$	NNR	$[0, +\infty)$	The amount of energy stored in the EES
23	$Ch_{s,t,n}$	NNR	$[0, +\infty)$	Input energy to the EES
24	$Dis_{s,t,n}$	NNR	$[0, +\infty)$	Output energy from the EES
25	$B_{s,t,n}^{Ch}$	Binary	$\{0,1\}$	Binary variable for controlling charge and discharge of the EES (if is 1, the system is charging)
26	$B_{s,t,n}^{Dis}$	Binary	$\{0,1\}$	Binary variable for controlling charge and discharge of the EES (if 1, the system is discharging)
Startup and shut down				
27	$v_{g,t,n}$	Binary	$\{0,1\}$	Binary variable for thermal unit startup (if 1, the unit is started at time t)
28	$w_{g,t,n}$	Binary	$\{0,1\}$	Binary variable for thermal unit shut down (if 1, the unit is turned off at time t)
Operation variables				

(continued)

Table 3.6 (continued)

Number	Decision variable	Domain	Bounds	Description
29	$u_{g,t,n}$	Binary	{0,1}	Binary variable for thermal unit operation (if 1, the unit is working)
30	$a_{g,t,n}$	Binary	{0,1}	Binary variable for reserve capacity when the system is off (1: unit does not participate in the reserve market)

NNR NonNegativeReals

Table 3.7 Optimization problem sets and their explanation

Number	Set	Description
1	G	Set of generators, $G = \{g_1, g_2, \ldots, g_m\}$
2	T	Set of time steps, $T = \{t_1, t_2, \ldots, t_m\}$
3	N	Set of nodes, $N = \{n_1, n_2, \ldots, n_m\}$
4	S	Set of storages, $S = \{s_1, s_2, \ldots, s_m\}$
5	R	Set of renewables, $R = \{r_1, r_2, \ldots, r_m\}$
6	L	Set of lines fitted on the nonlinear functions, $L = \{l_1, l_2, \ldots, l_m\}$

*m is the last members' index

References

1. Borgnakke C, Sonntag R (2009) Fundamentals of thermodynamics, 7th edn. Wiley, Amsterdam
2. Erichsen G, Zimmermann T, Kather A (2019) Effect of different interval lengths in a rolling horizon MILP unit commitment with non-linear control model for a small energy system. Energies 12:1003. https://doi.org/10.3390/en12061003
3. Cuviella C, Colmenar-Santos A, Borge-Diez D, Rosales E (2019) Sanitary-ware factories: heat recovery strategies to optimize energy and water consumption. Energy Proced 157:719–736. https://doi.org/10.1016/j.egypro.2018.11.238
4. Zivic Djurovic M, Milacic A, Krsulja M (2012) A simplified model of quadratic cost function for thermal generators. In: Proceedings of the 23rd DAAAM international symposium intelligent manufacturing automous, vol 1, pp 25–28. https://doi.org/10.2507/23rd.daaam.proceedings.006
5. Van Hoorebeeck L, Absil PA, Papavasiliou A (2020) Global solution of economic dispatch with valve point effects and transmission constraints. Electr Power Syst Res 189:106786. https://doi.org/10.1016/j.epsr.2020.106786
6. Zhang R, Zhou J, Mo L, Ouyang S, Liao X (2013) Economic environmental dispatch using an enhanced multi-objective cultural algorithm. Electr Power Syst Res 99:18–29. https://doi.org/10.1016/j.epsr.2013.01.010

7. Abido MA (2009) Multiobjective particle swarm optimization for environmental/economic dispatch problem. Electr Power Syst Res 79(7):1105–1113. https://doi.org/10.1016/j.epsr.2009.02.005
8. Talbi EH, Abaali L, Skouri R, El Moudden M (2020) Solution of economic and environmental power dispatch problem of an electrical power system using BFGS-AL algorithm. Proced Comput Sci 170:857–862. https://doi.org/10.1016/j.procs.2020.03.144
9. Zhang H, Vittal V, Heydt G, Quintero J (2012) A mixed-integer linear programming approach for multistage security-constrained transmission expansion planning. IEEE Trans Power Syst 27:1125–1133. https://doi.org/10.1109/TPWRS.2011.2178000
10. Ghadertootoonchi A, Moeini-Aghtaie M, Davoudi M (2023) A hybrid linear programming-reinforcement learning method for optimal energy hub management. IEEE Trans Smart Grid 14(1):157–166. https://doi.org/10.1109/TSG.2022.3197458
11. Maleki A, Askarzadeh A (2014) Optimal sizing of a PV/wind/diesel system with battery storage for electrification to an off-grid remote region: a case study of Rafsanjan, Iran. Sustain Energy Technol Assess 7:147–153. https://doi.org/10.1016/j.seta.2014.04.005
12. Faisal M, Hannan MA, Ker PJ, Hussain A, Mansor MB, Blaabjerg F (2018) Review of energy storage system technologies in microgrid applications: issues and challenges. IEEE Access 6:35143–35164. https://doi.org/10.1109/ACCESS.2018.2841407
13. Hedman KW, Ferris MC, O'Neill RP, Fisher EB, Oren SS (2010) Co-optimization of generation unit commitment and transmission switching with N − 1 reliability. IEEE Trans Power Syst 25(2):1052–1063. https://doi.org/10.1109/TPWRS.2009.2037232
14. Bai L, Li F, Jiang T, Jia H (2017) Robust scheduling for wind integrated energy systems considering gas pipeline and power transmission N–1 contingencies. IEEE Trans Power Syst 32(2):1582–1584. https://doi.org/10.1109/TPWRS.2016.2582684

Chapter 4
Pyomo Formulation of Electricity System Optimization Problem

Abstract Coding the energy system optimization problem has several stages including: (1) determining the input parameters and categorizing them into different groups such as global parameters, node-based parameters, and technical parameters; (2) Importing the essential libraries; (3) Defining sets and decision variables; (4) Defining the objective function and operational constraints, and (5) Using the numerical solver to find the optimal solution. These steps are introduced in this chapter a long with the utilized Python code.

4.1 Introduction

In this chapter, a case study is presented and solved using the mathematical formulation described in Chap. 3 and Pyomo commands explained in Chap. 2. This chapter starts with describing the modeling parameters and characteristics of thermal and renewable generators. After providing all the essential information, Pyomo formulation of the objective function and constraints are discussed and the model is solved. Finally, different options to manipulate and solve the optimal results are presented.

4.2 Input Parameters

There are seven types of inputs for the optimization problem. These inputs characterize the generators, storage units, electricity price and load as well as electricity transmission lines. This section describes all the seven input types and introduces their values used in the case study.

A. Ghadertootoonchi et al., *Energy System Modeling and Optimization*,
SpringerBriefs in Energy, https://doi.org/10.1007/978-3-031-65906-5_4

Table 4.1 Global parameters of the model

Parameter	Description	Equations	Case study value
EE^{max}	Maximum allowable excess electricity generation over the optimization horizon for all the nodes	(3.43)	500
LL^{max}	Maximum allowable unsatisfied electricity demand over the optimization horizon for all the nodes	(3.44)	500
Em^{max}	Maximum allowable greenhouse gas emission over the optimization horizon for all the nodes	(3.70)	50
ω_c	Weight of economical costs in the objective function	(3.9)	0.5
ω_e	Weight of environmental costs in the objective function	(3.9)	0.5
l_{max}	Number of lines fitted on the cost and emission curves of the thermal power plants for linearization	(3.20)	3
r_{max}	Maximum percentage of reserve surplus to demand	(3.69)	0.75

4.2.1 Global Parameters

Global parameters are applied to all the nodes and are not node-based. There are seven global parameters in total as shown in Table 4.1.

Since the optimization problem has two objectives (minimizing the economic and environmental costs), two weights are used to turn the problem into a single objective model where $\omega_c + \omega_e = 1$. In addition, the units of emission and electricity generation can be chosen arbitrarily but must be consistent across the model. Here the unit of emission is Ton and the unit of electricity is MW. The next section introduces the electricity price and load in different nodes.

4.2.2 Electricity Price and Load

The case study of this chapter consists of three nodes each with a different electricity load. The electricity price and reserve incentive are the same for the entire network (i.e., prices are similar for all the nodes). Figure 4.1 shows the electricity selling price (PP_t, see Eq. 3.17) and reserve incentive (RP_t, see Eq. 3.18).

As can be seen from Fig. 4.1, during peak hours, electricity price increases to encourage consumers reduce their consumption. This also increases the motivation to use storage units and participate in demand response and load shifting programs. In addition, it can be seen that during peak hours, reserve incentive is also increased meaning that the system's operator is willing to pay more to the generator units to minimize the load loss probability. The power load in each of the three nodes is shown in Fig. 4.2.

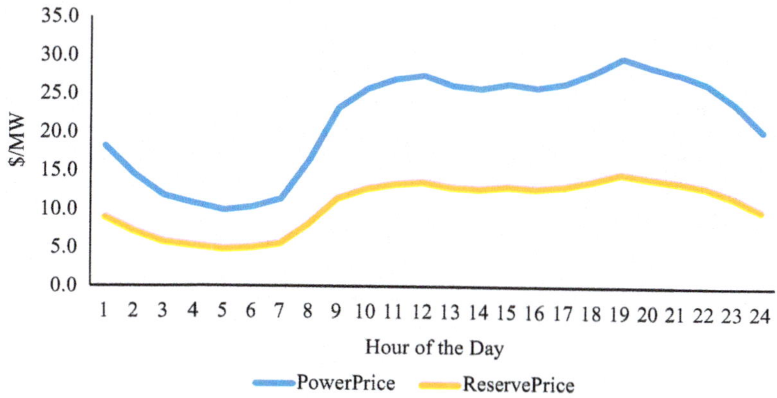

Fig. 4.1 Network electricity price and reserve incentive

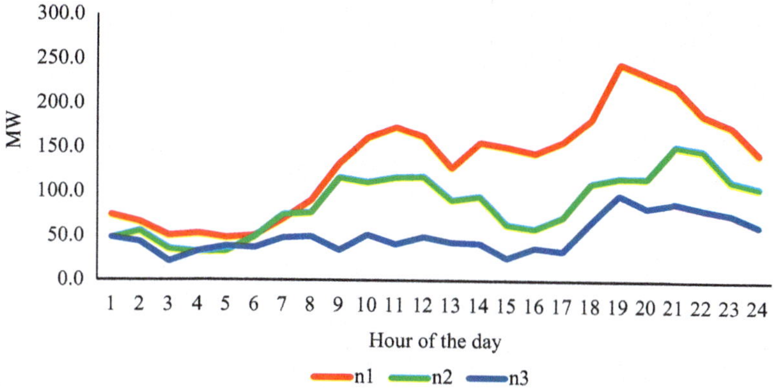

Fig. 4.2 Electricity load in three nodes of case study

The following section demonstrates the characteristics of storage units along with the thermal generation units.

4.2.3 Thermal Generators and Storage Units

This section presents the specifications of the thermal generators as well as the storage units. There are four types of generators and four types of electricity storage systems in total. Tables 4.2 and 4.3 show the information of thermal units.

The value of these parameters is adopted from the literature review. The parameters of the emission function are based on [1], and those of the cost function, ramp-down, and ramp-up are based on [2]. The value of startup and shut-down costs are attained

Table 4.2 Value of generator parameters

Generator	$MUT(h)$	$MDT(h)$	$RU\left(\frac{MW}{h}\right)$	$RD\left(\frac{MW}{h}\right)$	$P_{min}(MW)$	$P_{max}(MW)$	$SUC\left(\frac{\$}{start}\right)$	$SDC\left(\frac{\$}{shutdown}\right)$	$SUE\left(\frac{ton}{start}\right)$	$SDE\left(\frac{ton}{shutdown}\right)$
g_1	2	2	30	15	20	100	100	15	0.225	0.075
g_2	1	2	15	15	10	50	100	15	0.675	0.225
g_3	2	2	60	60	20	90	100	15	0.652	0.2175
g_4	2	1	15	15	15	60	100	15	0.675	0.225
Equations	(3.27)	(3.28)	(3.29)	(3.30)	(3.21), (3.22)	(3.21), (3.22)	(3.10)	(3.10)	(3.11)	(3.11)

Generator	$a\left(\frac{\$}{MW^2}\right)$	$b\left(\frac{\$}{MW}\right)$	$c(\$)$	$\gamma\left(\frac{ton}{MW^2}\right)$	$\beta\left(\frac{ton}{MW}\right)$	$\alpha(ton)$	$\mu(ton)$	$\sigma\left(\frac{1}{MW}\right)$
g_1	0.038	14	32	0.0000021	0.00016385	0.0069297	0.00012485	0.012
g_2	0.013	13.9	105	0.00002326	−0.0019762	0.1750028	0.000127375	0.01234
g_3	0.021	15.4	29	0.0000023	−0.0002556	0.0214478	0.00012735	0.01234
g_4	0.015	12.1	82	0.00002326	−0.00195115	0.1650028	0.000125815	0.01234
Equations	(3.12)	(3.12)	(3.12)	(3.13)	(3.13)	(3.13)	(3.13)	(3.13)

Table 4.3 Value of storage unit parameters

Storage	η_s^{ch}	η_s^{dis}	κ_s^1	κ_s^2	$Cap_s(MWh)$	DOD_s^{max}	$Ch_s^{max}(MW)$	$Dis_s^{max}(MW)$
s_1	0.98	0.98	0.9998	0	2	0.92	0.25	0.25
s_2	0.97	0.97	0.9998	0	4	0.9	0.2	0.2
s_3	0.96	0.96	0.9998	0	1	0.95	0.1	0.1
s_4	0.95	0.95	0.9998	0	5	0.88	0.5	0.5
Equations	(3.56)	(3.56)	(3.56)	(3.56)	(3.57)	(3.57)	(3.58)	(3.59)

from [3], and the startup and shut-down emissions from [4]. The parameters of the ESSs are obtained from [5–9]. It is considered that the initial and final level of energy for all the electricity storage units are equal to the minimum allowable amount which is calculated based on Eq. (3.57). The next section explains the renewable energy generation.

4.2.4 Renewable Generation

Renewable power generation is considered as an exogenous input for the optimization model. Total amount of renewable electricity for each node can be entered in a spreadsheet and later be used in the model based on the equations described in Chap. 3. Figure 4.3 shows the values considered in the case study.

One option to attain renewable electricity data is to use the "renewable ninja" website. The website gives information on solar and wind energy on an hourly basis for a year. The user can download the data and introduce it to the optimization model manually or directly use the renewable ninja API. The description of the latter is out of the scope of this book and interested readers are referred to the python code of the book available on GitHub. In addition, it is possible to perform robust optimization in

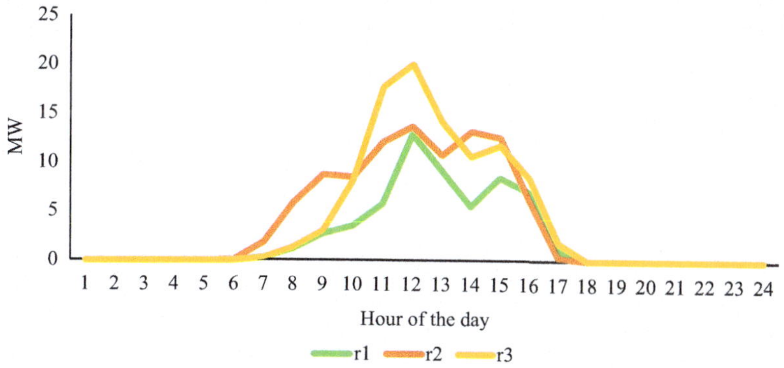

Fig. 4.3 Renewable power generation for three sources of r_1, r_2, r_3

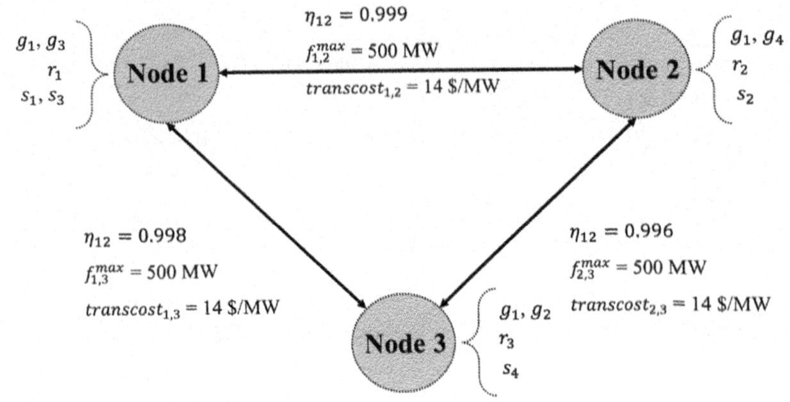

Fig. 4.4 Location of energy conversion units and map of transmission lines

presence of renewable energies. To do so, one must obtain solar electricity generation in the worst-case scenario and solve the model to see the network performance under such a condition. The following section demonstrates the distribution of thermal power plants, storage units, and renewable technologies.

4.2.5 *Power Conversion Units and Transmission Lines*

In order to solve the economic and environmental dispatch problem, the location of each of the energy conversion units must be determined. Additionally, these units must be connected to each other to form a network; Therefore, the characteristics of the transmission lines must also be known. This information is brought in Fig. 4.4.

As is clear from Fig. 4.4, there are three nodes in the optimization model connected by three transmission lines. The efficiency ($\eta_{n,n'}$), capacity ($f_{n,n'}^{max}$), and transmission costs ($TRC_{n,n'}$) are shown in Fig. 4.4 [10]. It should be noted that no transmission line outage is considered in this case ($Out_{t,n,n'} = 1\forall t \in T, n\&n' \in N$). Furthermore, the type of generators utilized in each node along with the considered storage and renewable technology are depicted in Fig. 4.4. Finally, value of node-based parameters is brought in the following section.

4.2.6 *Node-Based Parameters*

The value of these parameters is defined for each node as shown in Table 4.4. Please refer to the equations introduced in Chap. 3 for the explanation of these parameters.

Although the carbon price (EMC_n) has reached more than 100 Euros/Ton in the EU in February 2023 [11], the price is less in other places. Thus, in this study, it is

Table 4.4 Node-based parameters

Node	$Em_n^{max}(ton)$	$EMC_n\left(\frac{\$}{ton}\right)$	$EC_n\left(\frac{\$}{MW}\right)$	$LC_n\left(\frac{\$}{MW}\right)$	$EE_{n,hourly}^{max}(MW)$	$LL_{n,hourly}^{max}(MW)$	$EE_n^{max}(MW)$	LL_n^{max}
n_1	20	10	200	200	100	100	300	300
n_2	20	10	200	200	100	100	300	300
n_3	20	10	200	200	100	100	300	300
Equations	(3.71)	(3.11)	(3.14)	(3.15)	(3.39)	(3.40)	(3.41)	(3.42)

Node	ψ_{up}	ψ_{down}	$NZEC_n$
n_1	0.2	0.2	0.5
n_2	0.2	0.2	0.5
n_3	0.2	0.2	0.5
Equations	(3.46)	(3.46)	(3.53)

considered to be 10 $/Ton [12]. It should be mentioned that changing the $NZEC_n$ may result in model infeasibility. Finally, since there are a lot of parameters for the optimization problem, it is a good idea to develop a spreadsheet containing the value of these parameters. Doing so will increase the flexibility of modeling procedure and makes the sensitivity analysis more effective. Such a spreadsheet can be found on the GitHub page of the book.

4.3 Pyomo Formulation

This section discusses the Pyomo code of the optimization model presented in Chap. 3. The order of the python code boxes is aligned with the mathematical formulations of Chap. 3. The section is started with introducing the necessary libraries like NumPy, Pandas, and Pyomo. Then linearization is discussed and optimization variables along with sets are defined. Finally, the objective functions and constraints of the optimization problem are brought. At the end of this section, the reader has a complete python implementation of the economic and environmental optimization model. A Jupyter notebook version of the code is available in GitHub.

4.3.1 Libraries and Input Data

Three libraries are needed in order to develop a Python code for the optimization problem, as shown in Box 4.1. These are Pandas, NumPy, Pyomo. Pandas is used to read the input data from the excel spreadsheet, NumPy is mainly used for the linearization stage, and Pyomo is the main optimization library.

Box 4.1

```
1 from pyomo.opt import SolverStatus, TerminationCondition
2 import pyomo.environ as pyo
3 import pandas as pd
4 import numpy as np
```

In line 1, `SolverStatus` represents the state of the solver; If it is ok, then the solver is interpreted the problem correctly. On the other hand, `TerminationCondition` shows whether the solution is optimal or not. If the value is "optimal," then the global optimum is found. Possible values for solver status and termination condition can be found in Pyomo documentation [13]. Furthermore, to read the input data from the Excel file, one can use Pandas library as shown in Box 4.2.

Box 4.2

```
1 GenData = pd.read_excel(address + "Data.xlsx",header = 0,index_col = 0,sheet_name = "Generators")
2 GenNode = pd.read_excel(address + "Data.xlsx",header = 0,index_col = 0,sheet_name = "GenNodes")
3 G = list(GenData.index) # will be used as the set of generators
NumGen = len(G)
4 load = pd.read_excel(address + "Data.xlsx",header = 0,index_col = 0,sheet_name = "Loads")
5 T = list(load.index) # will be used as the set of time steps
6 N = list(load.columns) # will be used as the set of nodes
7 Horizon = len(T)
```

where address is the directory at which the Excel file is located. Generators' data is acquired from the Generators sheet which is shown is Fig. 4.5.

As shown in Box 4.2, the set of generators (G) is created using the indices of the input data which are {g1, g2, g3, g4}. Other characteristics like quadratic cost function coefficients and emission function coefficients are also provided in this sheet. These data will be used in the next section for linearization. Additionally, the electricity consumption of each node is provided in the "Loads" sheet as depicted in Fig. 4.6.

Electricity node is imported to the model using line 4 in Box 4.2 and the set of time steps and nodes are specified using the indices and column names (lines 5 and 6). Finally, the horizon of the optimization model (total number of time steps) is attained based on the number of time steps in line 7. Other data provided in the Excel file is imported using a similar process. Having the input data, one can linearize.

	MinOnTime	MinOffTime	RampUp	RampDown	pmin	pmax	StartUpCost	ShutDownCost
g1	2	2	30	15	20	100	100	15
g2	1	2	15	15	10	50	100	15
g3	2	2	60	60	20	90	100	15
g4	2	1	15	15	15	60	100	15

Fig. 4.5 Input data in the generators sheet (first 8 columns)

Fig. 4.6 Electricity load of each node (first 4 h)

time	n1	n2	n3
1	75.0	49.0	49.5
2	67.0	56.9	44.3
3	51.7	36.6	22.3
4	53.9	33.4	33.8

4.3.2 Linearization

GenData (Box 4.2), contains all the information related to the generator units including the quadratic cost and emission functions coefficients [Eqs. (3.6) and (3.7)] as well as minimum and maximum capacity of the generators. Therefore, one can create a list of cost/emission coefficients and P_{min}/P_{max} values as represented in Box 4.3.

Box 4.3

```
1 cost_coef = [list(GenData.loc[:,"c"]),
          list(GenData.loc[:,"b"]),
          list(GenData.loc[:,"a"])]

2 em_coef = [list(GenData.loc[:,"alpha"]),
         list(GenData.loc[:,"beta"]),
         list(GenData.loc[:,"gamma"]),
         list(GenData.loc[:,"eta"]),
         list(GenData.loc[:,"sigma"])] # emission coefficients

3 pmin = np.array(GenData.loc[:,"pmin"])
4 pmax = np.array(GenData.loc[:,"pmax"])
Results:
5 cost_coef = [[32, 105, 29, 82],
          [14.0, 13.9, 15.4, 12.1],
          [0.0382, 0.0127, 0.0212, 0.0148]]
6 pmin = [20, 10, 20, 15]
```

In the cost_coef list, c is located at index zero since it is the constant value or the coefficient of the term p^0. Similarly, b is the coefficient of p, and a is the coefficient of p^2. This is also the case for em_coef list; However, here there are two more coefficients that are related to the exponential term in Eq. (3.7). Finally, a list of minimum and maximum power generation capacities is created in lines 3 and 4. After running the code, the cost_coef list would be equal to that of line 5, where the first list (first line) in the nested list represents the c values for generator 1–4, the second list contains the b values, and the last one has the c values. Moreover, line 6 shows the minimum capacity of generator 1–4. Similar results would be observed for em_coef and P_{max}.

After creating these lists, the quadratic cost function and the quadratic-exponential emission functions should be defined based on Box 4.4. These two functions represent actual cost and emission of each generator given its power output.

Box 4.4

```
def GenCost(cost_coef,P):
    return (cost_coef[2]*np.power(P, 2) + cost_coef[1]*P +
cost_coef[0]).T
```

```
3 def GenEmission(emission_coef,P):
4 return (em_coef[2] *np.power(P,2) + em_coef[1] *P + em_coef[0]
+ em_coef[3] *np.exp(em_coef[4] *P)).T
Results:
5 P1 = np.array([25,15,25,20]).reshape(1,-1)
6 P2 = np.array([[20,10,20,15],
          [25,15,25,20]])
7 GenCost(cost_coef,P1) =
8 array([[405.875],
      [316.3575],
      [427.25 ],
      [329.92 ]])
9 GenCost(cost_coef,P2) =
10 array([[327.28 , 405.875],
      [245.27 , 316.3575],
      [345.48 , 427.25 ],
      [266.83 , 329.92 ]])
```

In the above functions, P is a NumPy array with shape of (n, 4). For instance, if one set of power output data (P1, line 5) is available, the operation cost of each generator would be similar to line 6, where the value in the first list represents the operational cost of the first generator, and the second value is for the second generator and so on. Nonetheless, if two sets of data are available, the operational cost for each of them would be equal to line 10. The next step is to specify the number of lines that must be fitted to the nonlinear cost and emission functions and finding their slope. This procedure is shown in Box 4.5.

Box 4.5

```
lines_num = int(Params.loc['lines_num','Val']) # number of
lines to fit on function
points_num = lines_num + 1
points = np.linspace(pmin,pmax,num = points_num)
costs = GenCost(cost_coef,points)
emissions = GenEmission(emission_coef,points)
points = points.T

CostSlope = np.zeros((NumGen, lines_num))
EmissionSlope = np.zeros((NumGen, lines_num))
for i in range(NumGen):
    for j in range(points_num-1):
    CostSlope[i,j] = (costs[i,j + 1] - costs[i,j])/(points[i,j
+ 1] - points[i,j])
    EmissionSlope[i,j] = (emissions[i,j + 1]-emissions[i,j])/
(points[i,j + 1]- points[i,j])

linear_range = []
for i in range(NumGen):
    linear_range.append(points[i,1] - points[i,0])
```

Cost Slope			
	1	2	3
g1	16.546667	18.584	20.621333
g2	14.323333	14.662	15.000667
g3	16.742667	17.732	18.721333
g4	12.766000	13.210	13.654000

Emission Slope			
	1	2	3
g1	0.000306	4.189411e-04	0.000532
g2	-0.001199	-5.783214e-04	0.000042
g3	-0.000108	5.086295e-07	0.000109
g4	-0.000902	-2.041803e-04	0.000494

Range	
	linear_range
g1	26.666667
g2	13.333333
g3	23.333333
g4	15.000000

Fig. 4.7 Results of the linearization process

First, the number of lines is determined in line 1. This parameter can be altered in the "Params" sheet of the Excel file. The number of points is one more than the number of lines. After finding the number of points, the value of the points is attained in line 3 using NumPy linspace. This function divides the distance between a minimum and a maximum point to sections with equal length. For instance, if one chooses to fit three lines for generator 4 with $P_{min} = 15$ and $P_{max} = 60$, the points will be [15, 30, 45, 60]. After finding the points, cost and emission values at those points are specified in lines 4 and 5. At the next step, the slopes of cost and emission functions are calculated in lines 7–12 based on Eq. (3.19). Finally, the distance between two consecutive points, or the len parameter in Fig. 3.8, is calculated in lines 13–15 using the distance between two adjacent points. The final results are shown in Fig. 4.7.

Figure 4.7 vividly shows that the slopes of cost and emission function increase as we move towards the maximum power output (from line 1 to line 3). Therefore, the more the power output of the generator, the more the operational costs and emissions. In addition, the range shows that considering three lines divided the power output of generator 1 to three sections with 26.67 MW capacity, while for generator 2 the range is 13.33 MW. After linearizing the cost and emission functions, the sets and variables of the optimization model should be defined. This is done in the next section.

4.3.3 *Sets and Variables Definition*

So far, the necessary libraries and data are imported to the model and nonlinear functions are linearized so that open-source linear programming solvers can be used to deal with the problem. This section defines all the required sets and decision variables. In Box 4.6, the model is initialized, and the sets (Table 3.7) are defined.

Box 4.6

```
model = pyo.ConcreteModel(name = "EcoEvnOpt")

# G  defined when reading generators' data (see Box 4.2)
# T  defined when reading load data (see Box 4.2)
# N  defined when reading load data (see Box 4.2)
```

```
# S  defined when reading storages' data
# R  defined when reading renewables data
2 L = list(range(1,lines_num + 1)) # lines fitted on the cost and
emission functions
```

The first line in Box 4.6 initializes the model and specifies its name. The comments mention that those sets are defined when reading the input data from the Excel file. Set of storages can be constructed using the "Storage" sheet indices, while the set of renewables can be constructed using the column names in the "RenNode" sheet. Finally, the set of lines fitted on the nonlinear functions is created using line 2. The DVs are categorized into eight groups and described in the following. Firstly, the variables related to the OF are introduced in Box 4.7 [see, Eq. (3.9)].

Box 4.7

```
model.SUSDC  = pyo.Var(within = pyo.NonNegativeReals)
model.SUSDEC = pyo.Var(within = pyo.NonNegativeReals)
model.OC     = pyo.Var(within = pyo.NonNegativeReals)
model.OEC     = pyo.Var(within = pyo.NonNegativeReals)
model.EEC     = pyo.Var(within = pyo.NonNegativeReals)
model.CEC     = pyo.Var(within = pyo.NonNegativeReals)
model.TC     = pyo.Var(within = pyo.NonNegativeReals)
model.PI     = pyo.Var(within = pyo.NonNegativeReals)
model.RI     = pyo.Var(within = pyo.NonNegativeReals)
```

All the variables are nonnegative. The naming pattern used in the code is exactly similar to the formulation provided in Chap. 3. The next variables are related to power generation and reserve capacity as shown in Box 4.8.

Box 4.8

```
1 model.pl = pyo.Var(G,T,L,N,within = pyo.NonNegativeReals)
2 model.p    = pyo.Var(G,T,N,   within = pyo.NonNegativeReals,
initialize = 0)
3 model.r   = pyo.Var(G,T,N,  within = pyo.NonNegativeReals)
4 for g in G:
5     for t in T:
6         for n in N:
7             if GenNode.loc[n,g] ! = 1:
8                 model.r[g,t,n].fix(0)
```

Variables defined in lines 1–3 are similar to variable number 10, 11, and 12 brought in Table 3.6. The for loop is added so that if generator g is not in node n, its reserve capacity is fixed to zero. Similar concept is defined for power output using the binary variables in the following sections to show that the same goal is attainable using different formulations. The flow of generated power between the nodes is defined based on Box 4.9.

Box 4.9

```
1 model.f  = pyo.Var(T,N,N,within = pyo.NonNegativeReals)
2 for t in T:
3     for n in N:
4         for nn in N:
5             if n == nn: # one cannot send energy from n to n (itself)
6                 model.f[t,n,nn].fix(0)
7 model.fb  = pyo.Var(T,N,N,within = pyo.Binary)
```

Line 1 in Box 4.9 defines the amount of power flow from one node to another. However, a node cannot send electricity to itself; The for loop prevents such condition. The binary variable in line 7 is similar to $\chi_{t,n,n'}$ in Eqs. (3.47) and (3.48). The amount of excess and curtailed power generation is specified using Box 4.10.

Box 4.10

```
1 model.ren_consumption = pyo.Var(T,N, within =
pyo.NonNegativeReals)
2 model.ren_excess = pyo.Var(T,N, within =
pyo.NonNegativeReals)

3 model.thermal_consumption = pyo.Var(T,N, within =
pyo.NonNegativeReals)
4 model.thermal_excess = pyo.Var(T,N, within =
pyo.NonNegativeReals)

5 model.EE = pyo.Var(T,N, within = pyo.NonNegativeReals)
6 model.CE = pyo.Var(T,N, within = pyo.NonNegativeReals)
```

Lines 1–4 in Box 4.10 divide the renewable and thermal power to two parts [see Eqs. (3.34) and (3.35)]. Line 5 and 6 define the excess and curtailed electricity at each time step for each node (variables 19 and 20 in Table 3.6). Another variable related to the electricity balance constraint is the demand response brought in Box 4.11.

Box 4.11

```
def demand_response_bounds(model,t,n):
    return( load.loc[t,n]*(1-NodeData.loc[n,'DRPlb']),
        load.loc[t,n]*(1 + NodeData.loc[n,'DRPub']))
model.D = pyo.Var(T,N, within = pyo.NonNegativeReals, bounds
= demand_response_bounds)
```

Box 4.11 shows that the lower and upper bound of the demand is specified using node-based demand response percentage which is an input parameter in the "Node-Data" sheet of the Excel file. Moreover, the GHG emission resulting from the operation of thermal units is represented in Box 4.12, and the DVs related to the EES units are brought in Box 4.13.

Box 4.12

```
model.emsn  = pyo.Var(G,T,N,within = pyo.NonNegativeReals)
```

Box 4.13

```
1 model.SLE   = pyo.Var(T,N,S, within = pyo.NonNegativeReals) #
level of energy
2 model.SCH   = pyo.Var(T,N,S, within = pyo.NonNegativeReals) #
charging input
3 model.SDIS = pyo.Var(T,N,S, within = pyo.NonNegativeReals) #
discharging output
4 model.SCHB = pyo.Var(T,N,S, within = pyo.Binary)
5 model.SDISB = pyo.Var(T,N,S, within = pyo.Binary)
6 for t in T:
7     for n in N:
8         for s in S:
9           #if storage s does not exist in node n, SLE, SCH, and SDIS
must be zero
10            if StorNodes.loc[n,s] = = 0:
11                model.SLE[t,n,s].fix(0)
12                model.SCH[t,n,s].fix(0)
13                model.SDIS[t,n,s].fix(0)
14                model.SCHB[t,n,s].fix(0)
15                model.SDISB[t,n,s].fix(0)
16                else:
17                    if t = = Horizon:
18                model.SLE[t,n,s].fix(Storage.loc
                    [s,'FinalEnergy'])
```

The variables defined in lines 1–5 of Box 4.13 follow a similar order to those presented in Table 3.6 (EES system section). The loop ensures that if the storage type s is not in node n, all the related variables are fixed to zero (based on the data in the "StoreNodes" sheet). In addition, at the final time step, the amount of energy stored in the storage unit is fixed to a predefined value called "Final Energy." This parameter is specified by the user in "Storage" sheet. Finally, the binary variable related to start up and shut down of the system as well as its operational status are defined in Box 4.14.

Box 4.14

```
1 model.v  = pyo.Var(G,T,N,  within = pyo.Binary, initialize = 0)
2 model.w  = pyo.Var(G,T,N,  within = pyo.Binary, initialize = 0)
3 model.u  = pyo.Var(G,T,N,  within = pyo.Binary, initialize = 0)
4 for g in G:
5     for t in T:
6         for n in N:
7             if GenNode.loc[n,g] == 0:
8                 model.u[g,t,n].fix(0)
9 model.a = pyo.Var(G,T,N,  within = pyo.Binary)
```

Variables defined in lines 1–3 and 9 are in similar order to variables 27–30 in Table 3.6. u is initialized with zero meaning that at the beginning, the generators are off. The loop states that if generator g is not in node n, its operation status should be fixed to zero (i.e., $P_{g,n,t} = 0 \forall t$).

Now that all the DVs are defined, the next step is defining the OF. The OF consists of 9 parts that are discussed in the next section.

4.3.4 *Objective Function*

As mentioned in Eq. (3.9), the OF has nine parts which are related to costs and income of the thermal units' operator. Pyomo formulation of these 9 costs and incomes are brought in the section. Firstly, the start up and shut down costs are brought in Box 4.15.

Box 4.15

```
def SUSD_Cost_rule(model):
    return model.SUSDC == sum( GenData.loc[g, 'StartUpCost']*model.v[g,t,n] +
                         GenData.loc[g,'ShutDownCost']*model.w[g,t,n]
                            for g in G for t in T for n in N)
model.SUSD_Cost = pyo.Constraint(rule = SUSD_Cost_rule)
```

The start up and shut down costs are provided by the user in the "Generators" sheet of the Excel file. Likewise, the start up and shut down emissions can be calculated based on Box 4.16.

Box 4.16

```
def SUSD_Emission_Cost_rule(model):
    return model.SUSDEC == sum( NodeData.loc[n,'EmissionCost']*
            (GenData.loc[g,'StartUpEmission']*model.v[g,t,n]
            +
            GenData.loc[g,'ShutDownEmission']*model.w[g,t,n])
                for n in N for g in G for t in T)
model.SUSD_Emission_Cost = pyo.Constraint(rule = SUSD_Emission_Cost_rule)
```

Similar to the start up and shut down costs, the emissions are also imported from the "Generators" sheet. The operational cost of the thermal units is divided into two parts. First, the fixed cost which should be paid if the generator is on regardless of the amount of power generation. This cost can be calculated using the quadratic cost function and the minimum power output. Second, the variable cost which is computed based on Eq. (3.23) (for further explanation, see Fig. 3.8).

Box 4.17

```
def Operation_Cost_rule(model):
    fixed_OC    = sum( GenData.loc[g,'MinCost']*model.u[g,t,n]
                  for g in G for t in T for n in N)
    variable_OC = sum( CostSlope.loc[g,l]*model.pl[g,t,l,n]
                  for g in G for t in T for l in L for n in N)
    return model.OC == fixed_OC + variable_OC
model.Operation_Cost = pyo.Constraint(rule = Operation_Cost_rule)
```

Moreover, the operational emission cost is calculated based on the code provided in Box 4.18.

Box 4.18

```
def Operation_emission_cost_rule(model):
    return model.OEC == sum (NodeData.loc[n,'EmissionCost']*model.emsn[g,t,n]
                  for g in G for t in T for n in N)
model.Operation_emission_cost = pyo.Constraint(rule = Operation_emission_cost_rule)
```

The emission cost in node n should be provided by the user in the "NodeData" sheet. Likewise, the penalties for excess power generation and load loss are imported from the "NodeData" sheet as shown in Box 4.19.

Box 4.19

```
def Excess_electricity_cost_rule(model):
    return model.EEC == sum( NodeData.loc[n,'excess_ele_
cost']*model.EE[t,n]
                          for t in T for n in N)
model.Excess_electricity_cost = pyo.Constraint(rule = Excess_
electricity_cost_rule)

def Curtailed_electricity_cost_rule(model):
    return model.CEC == sum( NodeData.loc[n,'curtailed_ele_
cost']*model.CE[t,n]
                          for t in T for n in N)
model.Curtailed_electricity_cost = pyo.Constraint(rule =
Curtailed_electricity_cost_rule)
```

The cost of transmitting electricity between nodes is another component of the OF, as defined in Box 4.20.

Box 4.20

```
def Transmission_cost_rule(model):
  return model.TC == sum( TransCost.loc[n,nn]*model.f[t,n,nn]
                          for t in T for n in N for nn in N)
model.Transmission_cost = pyo.Constraint(rule = Transmission_
cost_rule)
```

The data for the transmission costs between a pair of nodes should be specified in the "TransCost" sheet. The remaining two boxes are related to the income of the thermal units' operator. The reserve income is presented in Box 2.21.

Box 4.21

```
def Reserve_market_rule(model):
   return model.RI == sum( Market.loc[t,
'ReservePrice']*model.r[g,t,n]
                          for g in G for t in T for n in N)
model.Reserve_market = pyo.Constraint(rule = Reserve_market_
rule)
```

The incentive price to encourage the thermal units to participate in the reserve market should be determined by the user in the "Market" sheet for each time step. Also, the income from providing power is based on Box 4.22.

Box 4.22

```
def Power_market_rule(model):
    generation_income = sum(Market.loc[t,'PowerPrice']*
                    (model.thermal_consumption[t,n] + model.ren_
consumption[t,n])
                    for t in T for n in N)
    discharge_income = sum(Market.loc
[t,'PowerPrice']*model.SDIS[t,n,s]
                              for t in T for n in N for s in S)
    return model.PI == generation_income + discharge_income
model.Power_market = pyo.Constraint(rule = Power_market_rule)
```

where the power price is imported from the "Market" sheet. As can be seen, the total amount of power sold is equal to the total amount of power consumed [see, Eq. (3.33)] which consists of power generated by thermal units, renewables, and discharged from the EES units. Finally, the objective function of the problem is described in Box 4.23.

Box 4.23

```
def obj_rule(model):
    return Params.loc['cost_weight','Val']*
      (model.SUSDC + model.OC + model.EEC + model.CEC + model.TC-
model.RI-model.PI) +
           Params.loc['emission_weight','Val']*(model.SUSDEC +
model.OEC)
model.obj = pyo.Objective(rule = obj_rule, sense =
pyo.minimize)
```

As can be seen, the economic costs are multiplied by the cost weight (i.e., ω_c) while the emission costs are multiplied by the emission weight (i.e., ω_e). The weights are imported from the "Params" sheet. The following sections discuss the Pyomo formulation of the constraints.

4.3.5 Start up and Shut Down Constraints

Start up and shut down constraints change the values of v (start up binary variable) and w (shut down binary variable) as discussed in Sect. 3.5.1. Python code of these two constraints is brought in Box 4.24.

Box 4.24

```
1 def SU_rule(model,g,t,n):
2    if t > 1:
3            return model.u[g,t,n]  -  model.u[g,t-1,n]  <  =
model.v[g,t,n]
4    else:
5        return model.u[g,t,n] < = model.v[g,t,n]
6 model.SU = pyo.Constraint(G,T,N, rule = SU_rule)

7 def SD_rule(model,g,t,n):
8    if t > 1:
9            return model.u[g,t-1,n]  -  model.u[g,t,n]  <  =
model.w[g,t,n]
10   else:
11       return pyo.Constraint.Skip
12 model.SD = pyo.Constraint(G,T,N, rule = SD_rule)
```

The first 6 lines show the start up constraint. The equation is divided into two parts based on the time step. For the initial time step (i.e., $t = 1$), v must be greater than or equal to u so that of the generator must operate ($u = 1$), then it should be started ($v = 1$). For the remaining time steps, the formulation is similar to Eq. (3.25). Likewise, shut down constraint's code is shown in lines 7–12. The only difference is that there is no constraint for $t = 1$. This is because the initial value of u is zero (see Box 4.14), therefore, one cannot shut down a generator that is already off. Python code for minimum up and down time constraints is brought in the next section.

4.3.6 *Minimum up Time and Minimum Down Time Constraints*

Following the mathematical formulation shown in Sect. 3.5.2, the Python code for minimum up and down time constraints are discussed in Box 4.25. In order to implement these constraints, one must first check whether min up or down time exceeds the optimization horizon or not and choose the time step which is closer to the current one (i.e., choose between *Horizon* and $t + MUT - 1$). Then, a new set of time steps should be created during which the generator must be on or off.

Box 4.25

```
def min_up_time_rule(model,g,t,n):
   alpha = min(Horizon,t + GenData.loc[g,'MinOnTime']-1)
   TT = list(range(t,alpha + 1))
          return  sum(  model.u[g,tt,n]  for  tt  in  TT)  >  =
len(TT)*model.v[g,t,n]
```

```
5 model.MUT = pyo.Constraint(G,T,N, rule = min_up_time_rule)

6 def min_down_time_rule(model,g,t,n):
7     alpha = min(Horizon,t + GenData.loc[g,'MinOffTime']-1)
8     TT = list(range(t,alpha + 1))
9           return  sum(  1-model.u[g,tt,n]  for  tt  in  TT)  >   =
len(TT)*model.w[g,t,n]
10 model.MDT = pyo.Constraint(G,T,N, rule = min_down_time_rule)
```

Alpha in lines 2 and 7 represents the choice between *Horizon* and $t + MUT - 1$. Then, the new set of time steps (*TT*) is created and the generator is forced to be on or off during *TT* only if it was turned on ($v = 1$) or off ($w = 1$). Another operational constraint for the thermal units is ramp up and down which is brought in the following.

4.3.7 Ramp up and Ramp Down Constraints

Mathematical equation of ramp up and down constraints is discussed in Sect. 3.5.4 and the code is shown in Box 4.26. Similar to the start up and shut down formulation, there are different definitions for the initial time step and other time steps.

Box 4.26

```
def RU_rule(model,g,t,n):
    if t > 1:
        return model.p[g,t,n] - model.p[g,t-1,n] < =
                    GenData.loc[g,'RampUp']*(1 + model.u[g,t-1,n]-
model.u[g,t,n]) +
            GenData.loc[g,'pmin']*(2-model.u[g,t,n]-model.u[g,t-
1,n])
    else:
                                            return    model.p[g,t,n]    =   =
model.u[g,t,n]*GenData.loc[g,'pmin']
model.RU = pyo.Constraint(G,T,N, rule = RU_rule)

def RD_rule(model,g,t,n):
    if t > 1:
        return model.p[g,t-1,n] - model.p[g,t,n] < =
                  GenData.loc[g,'RampDown']*(1-model.u[g,t-1,n] +
model.u[g,t,n]) +
            GenData.loc[g,'pmin']*(2-model.u[g,t,n]-model.u[g,t-
1,n])
    else:
        return pyo.Constraint.Skip
model.RD = pyo.Constraint(G,T,N, rule = RD_rule)
```

Based on the ramp up constraint, if $t = 1$ (line 7) and the model decides to use the generator, then the power output would be equal to P_{min}. However, there is no constraint for $t = 1$ (line 15) in the ramp down constraint since u is initialized with zero (i.e., power output cannot be reduced further since the generator is off). Controlling power output so that there is no excess electricity generation or load loss is crucial for the desired performance of the network. Such a matter is discussed in the next section.

4.3.8 Excess Electricity and Curtailment Constraints

As mentioned in Sect. 3.5.6, there are six constraints related to excess power generation and load loss which are shown in Box 4.27. These constraints impose an upper bound on these two values in three levels. Firstly, hourly excess and curtailed electricity at each node should be controlled. This goal is achieved using the first two constraints (lines 1–6). Secondly, total excess and curtailed electricity for every node should be smaller than the user-defined upper bounds (second two constraints, lines 7–12). Lastly, total excess and curtailed electricity for the entire network during the optimization horizon must be less than the upper bound (the last two constraints, lines 13–18).

Box 4.27

```
1 def hourly_excess_ele_ub_node_rule(model,t,n):
2   return model.EE[t,n] < =NodeData.loc[n,'hourly_excess_ele_ub']
3 model.hourly_excess_ele_ub_node = pyo.Constraint(T,N, rule =
                                    hourly_excess_ele_ub_node_rule)

4 def hourly_curtailed_ele_ub_node_rule(model,t,n):
5   return model.CE[t,n] < =NodeData.loc[n,'hourly_curtailed_ele_ub']
6 model.hourly_curtailed_ele_ub_node = pyo.Constraint(T,N, rule =
                                    hourly_curtailed_ele_ub_node_rule)

7 def total_excess_ele_ub_node_rule(model,n):
8   return sum(model.EE[t,n] for t in T) < =NodeData.loc[n,'tot_excess_ele_ub']
9 model.total_excess_ele_ub_node = pyo.Constraint(N, rule =
                                    total_excess_ele_ub_node_rule)

10 def total_curtailed_ele_ub_node_rule(model,n):
11   return sum(model.CE[t,n] for t in T) < = NodeData.loc[n,'tot_curtailed_ele_ub']
12 model.total_curtailed_ele_ub_node = pyo.Constraint(N, rule =
```

```
                                                    total_curtailed_ele_ub_node_
rule)
def total_excess_ele_ub_rule(model):
    return sum(model.EE[t,n] for t in T for n in N) < =
        Params.loc['total_excess_ele_ub','Val']
model.total_excess_ele_ub = pyo.Constraint(rule = total_
excess_ele_ub_rule)
def total_curtailed_ele_ub_rule(model):
    return sum(model.CE[t,n] for t in T for n in N) < =
        Params.loc['total_curtailed_ele_ub','Val']
model.total_curtailed_ele_ub = pyo.Constraint(rule = total_
curtailed_ele_ub_rule)
```

The upper bound values for the first four constraints are determined by the modeler by changing the respective parameters in the "NodeData" sheet; Whereas, the last two upper bounds which are effective for the entire network can be adjusted from the "Params" sheet of the Excel input file. The following section represents the Python code for the electricity balance constraint.

4.3.9 Electricity Balance Constraints

In order to implement the power generation balance constraint, one must first define the demand response equation and determine the amounts of thermal and renewable-based power consumed. The code provided in Box 4.28 specifies these values.

Box 4.28

```
def demand_response_rule(model,n):
    return sum( model.D[t,n] for t in T) = = sum( load.loc[t,n]
for t in T)
model.demand_response = pyo.Constraint(N, rule = demand_
response_rule)
def thermal_generation_excess_rule(model,t,n):
    return sum( model.p[g,t,n] for g in G) = = model.thermal_
excess[t,n] +
                                   model.thermal_consumption[t,n]
model.thermal_generation_excess = pyo.Constraint(T,N, rule =
                                   thermal_generation_excess_rule)
def renewable_excess_rule(model,t,n):
    return sum(RenNode.loc[n,r]*RenGen.loc[t,r] for r in R) = =
        model.ren_excess[t,n] + model.ren_consumption[t,n]
model.renewable_excess = pyo.Constraint(T,N, rule = renew-
able_excess_rule)
```

```
10 def excess_ele_rule(model,t,n):
11     return model.EE[t,n] == model.ren_excess[t,n] + model.thermal_excess[t,n]
12 model.excess_ele = pyo.Constraint(T,N, rule = excess_ele_rule)
13 def demand_rule(model,t,n):
14     transaction = sum(TransEff.loc[nn,n]*model.f[t,nn,n]-model.f[t,n,nn] for nn in N)
15     storage  = sum(model.SDIS[t,n,s] - model.SCH[t,n,s] for s in S)
16     model.thermal_consumption[t,n] + model.ren_consumption[t,n] + storage + transaction -
        model.D[t,n] + model.CE[t,n] == 0
17 model.demand = pyo.Constraint(T,N, rule = demand_rule)
```

The first three lines are the demand response constraint which is discussed in Sect. 3.5.7. The second one (lines 4–6) divides the thermal power generation to two parts of excess and consumed. Renewable power is divided into similar parts using constraint three (lines 7–9). The amount of excess generation is calculated using the fourth constraint in lines 10–12. Finally, the balance equation is represented by the last constraint in lines 13–18. To define the balance constraint, first the net electricity transaction between one node and the others is calculated (line 14). Then, the net effect of storage units (produced or consumer) is computed (line 15). Lastly, the balance equality constraint is defined in line 16. Mathematical formulation of these equations is brought in Sect. 3.5.5.

4.3.10 Power Generation Constraints

To calculate the power output of a generator at a given time, follow the procedure described in Fig. 3.8, where the power output is determined as the sum of its linear components. In addition, if the generator is operating (i.e., $u = 1$), then the linear power outputs must be less than or equal to the length value which is calculated using Eq. (3.20) in Box 4.5. These two constraints are brought in Box 4.29.

Box 4.29

```
def linear_power_rule(model,g,t,n):
    return model.p[g,t,n] == GenData.loc[g,'pmin']*model.u[g,t,n] +
                    sum( model.pl[g,t,l,n] for l in L)
model.linear_power = pyo.Constraint(G,T,N, rule = linear_power_rule)
```

```
4 def linear_power_ub_rule(model,g,t,l,n):
5     return model.pl[g,t,l,n] <= linear_range.loc[g,'linear_range']*model.u[g,t,n]
6 model.linear_power_ub = pyo.Constraint(G,T,L,N, rule = linear_power_ub_rule)
```

Where lines 1–3 compute the power output and lines 4–6 impose the upper bound on the linear components. The next section discusses the integration of GHG emission equations into the optimization model.

4.3.11 *Emission Constraints*

Similar to operational costs, operational emissions are divided into two parts: fixed emissions and variable emissions. The fixed part is calculated by replacing P in the quadratic-exponential emission equation with P_{min}. It is an input parameter imported from the "Generators" sheet of the Excel file. On the other hand, the variable emission is calculated using the slope of lines fitted on the nonlinear emission function and the linear component of the power generation. This process is shown in the Box 4.30 (lines 1–5).

Box 4.30

```
1 def linear_emission_rule(model,g,t,n):
2     fixed_emsn = GenData.loc[g,'MinEmsn']*model.u[g,t,n]
3     variable_emsn = sum( EmissionSlope.loc[g,l]*model.pl[g,t,l,n] for l in L)
4     return model.emsn[g,t,n] == fixed_emsn + variable_emsn
5 model.linear_emission = pyo.Constraint(G,T,N, rule = linear_emission_rule)

6 def total_node_emission_ub(model,n):
7     return sum(model.emsn[g,t,n] for g in G for t in T) <= NodeData.loc[n,'EmissionUb']
8 model.total_node_emission = pyo.Constraint(N, rule = total_node_emission_ub)

9 def total_emission_ub_rule(model):
10    return sum( model.emsn[g,t,n] for g in G for t in T for n in N) <=
11           Params.loc['total_emission_ub','Val']
12 model.total_emission_ub = pyo.Constraint(rule = total_emission_ub_rule)
```

In Sect. 3.5.12, two upper bounds on emissions are detailed. The node-based upper bound is formulated using lines 6–8, while the network upper bound is defined in lines 9–12. Additionally, the next section will cover the power transmission constraints.

4.3.12 *Transmission Constraints*

The constraints related to electricity transmission are discussed in this section. Since energy self-sufficiency is closely related to the power transmission between the nodes, it is also discussed here. Box 4.31 shows the Python implementation of these equations.

Box 4.31

```
1 def net_zero_energy_rule(model,n):
2     return sum( model.f[t,n,nn] for nn in N for t in T) > =
          NodeData.loc[n, 'NetZeroEnCoef']*
          sum(model.f[t,nn,n]*TransEff.loc[nn,n] for nn in N for t
in T)
3 model.net_zero_energy = pyo.Constraint(N,rule = net_zero_
energy_rule)

4 def trans_ub_first_rule(model,t,n,nn):
5   return model.f[t,n,nn] < = TransUb.loc[n,nn]*model.fb[t,n,nn]
6 model.trans_ub_first = pyo.Constraint(T,N,N,rule = trans_ub_
first_rule)

7 def trans_ub_second_rule(model,t,n,nn):
8            return model.f[t,nn,n]  <   =  TransUb.loc[nn,n]*(1-
model.fb[t,n,nn])
9 model.trans_ub_second = pyo.Constraint(T,N,N,rule = trans_ub_
second_rule)
```

Pyomo code of the net-zero energy constraint is presented in lines 1–3 which is identical to the formulation introduced in Sect. 3.5.9. The net-zero energy coefficient (*NZEC*) is an input parameter imported from the "NodeData" sheet. Likewise, transmission line efficiency is provided in the "TransEff" sheet. In addition, the constraints that control the flow direction (see Sect. 3.5.8) are brought in lines 4–9.

4.3.13 *Storage Constraints*

There are three main types of constraints related to EES systems: the dynamic equation for calculating the amount of energy stored in the system, upper and lower bounds on the amount of stored energy, and upper bounds on the charging and discharging power. These constraints are detailed in Box 4.32.

Box 4.32

```
def Storage_level_of_energy_rule(model,t,n,s):
    if t > 1:
        return model.SLE[t,n,s] ==
Storage.loc[s,'SelfDisFactor']*model.SLE[t-1,n,s] +
                        Storage.loc[s,'ChEff']*model.SCH[t,n,s]-
                  (1/Storage.loc[s,'DisEff'])*model.SDIS[t,n,s]-
                 Storage.loc[s,'SelfDisConst']*StorNodes.loc[n,s]
    else:
        return model.SLE[t,n,s] ==
Storage.loc[s,'SelfDisFactor']*
        Storage.loc[s,'InitEnergy']*StorNodes.loc[n,s] +
       Storage.loc[s,'ChEff']*model.SCH[t,n,s]-
         (1/Storage.loc[s,'DisEff'])*model.SDIS[t,n,s]-
        Storage.loc[s,'SelfDisConst']*StorNodes.loc[n,s]
model.Storage_level_of_energy = pyo.Constraint(T,N,S, rule =
                                   Storage_level_of_energy_rule)
def Storage_level_of_energy_ub_rule(model,t,n,s):
    return model.SLE[t,n,s] <= Storage.loc[s,'SLEub']
model.Storage_level_of_energy_ub = pyo.Constraint(T,N,S,
rule =
                                Storage_level_of_energy_ub_rule)
def Storage_level_of_energy_lb_rule(model,t,n,s):
    return Storage.loc[s,'SLElb']*StorNodes.loc[n,s] <=
model.SLE[t,n,s]
model.Storage_level_of_energy_lb = pyo.Constraint(T,N,S,
rule =
                                Storage_level_of_energy_lb_rule)
def Storage_charging_cap_ub_rule(model,t,n,s):
    return model.SCH[t,n,s] <= Storage.loc[s,'ChCap']*m
odel.SCHB[t,n,s]
model.Storage_charging_cap_ub = pyo.Constraint(T,N,S, rule =
                                 Storage_charging_cap_ub_rule)
def Storage_discharging_cap_ub_rule(model,t,n,s):
    return model.SDIS[t,n,s] <= Storage.loc[s,'DisCap']*m
odel.SDISB[t,n,s]
model.Storage_discharging_cap_ub = pyo.Constraint(T,N,S,
rule =
                               Storage_discharging_cap_ub_rule)
def storage_charge_control_rule(model,t,n,s):
    return model.SDISB[t,n,s] + model.SCHB[t,n,s] <= 1
model.storage_charge_control = pyo.Constraint(T,N,S, rule =
                                 storage_charge_control_rule)
```

The first constraint (lines 1–6) divides the level of stored energy into two parts; For the initial time step, the "InitEnergy" parameter is used which is specified by the user in the "Storage" sheet. This value is the available energy in the EES system at the beginning. For the other time steps, Python implementation of Eq. (3.56) is used. In addition, based on Eq. (3.57), there is an upper bound and a lower bound for the level of stored energy which is implemented in lines 7–12. Finally, Eqs. (3.58)–(3.68) is represented by lines 13–21, respectively.

4.3.14 Reliability ($N - 1$ Contingency) Constraints

As detailed in Sect. 3.5.11, there are six constraints pertaining to the network's reliability. These constraints are presented in Box 4.33, with their descriptions provided in the following.

Box 4.33

```
def gen_contingency_reserve_ub_first_rule(model,g,t,n):
    return model.r[g,t,n] < = GenData.loc[g,'RampUp']*model.u[g,t,n] +
        GenData.loc[g,'pmin']*(1-model.u[g,t,n]-model.a[g,t,n])
model.gen_contingency_reserve_ub_first = pyo.Constraint(G,T,N, rule =
        gen_contingency_reserve_ub_first_rule)

def gen_contingency_reserve_lb_rule(model,g,t,n):
    return model.r[g,t,n] > = GenData.loc[g,'pmin']*(1-model.u[g,t,n]-
        model.a[g,t,n])
model.gen_contingency_reserve_lb = pyo.Constraint(G,T,N, rule =
        gen_contingency_reserve_lb_rule)

def gen_contingency_off_rule(model,g,t,n):
    return model.a[g,t,n] < = 1 - model.u[g,t,n]
model.gen_contingency_off = pyo.Constraint(G,T,N, rule = gen_contingency_off_rule)

def gen_contingency_reserve_ub_second_rule(model,g,t,n):
    return model.r[g,t,n] < = GenData.loc[g,'pmax'] - model.p[g,t,n]
model.gen_contingency_reserve_ub_second = pyo.Constraint(G,T,N, rule =
        gen_contingency_reserve_ub_second_rule)

def gen_contingency_lb_rule(model,g,t,n):
```

```
14    return sum(model.r[gg,t,nn] for gg in G for nn in N if gg! =
g or nn! = n) > =
             model.p[g,t,n]
15 model.gen_contingency_lb = pyo.Constraint(G,T,N, rule = gen_
contingency_lb_rule)

16 def gen_contingency_ub_rule(model,g,t,n):
17    return sum(model.r[gg,t,nn] for gg in G for nn in N if gg! =
g or nn! = n) < =
            model.p[g,t,n]*(1 + Params.loc['reserve_ub','Val'])
18 model.gen_contingency_ub = pyo.Constraint(G,T,N, rule =
                              gen_contingency_ub_rule)
```

The first equation (lines 1–3) imposes an upper bound on the reserve considering the ramp up rate and minimum power output of the generator [see Eq. (3.64)]. The second constraint (lines 4–6) defines a lower bound for the reserve considering P_{min} similar to Eq. (3.65). Also, the equation defining the relationship between the binary variables a and u is represented in lines 7–9 similar to Eq. (3.66). There is another upper bound for the offered reserve which is shown in lines 10–12 Eq. (3.67). Finally, the total amount of offered reserve should be limited by an upper and a lower bound as stated in Eqs. (3.68) and (3.69). This is done using lines 13–18.

N − 1 contingency is the last set of constraints used in the optimization model. In the next section, the code required to solve the model is introduced.

4.3.15 Solving the Model

The code required to solve the mode is similar to those mentioned in Chap. 2. However, here instead of using `check_optimal_termination` method, solver status and termination conditions are checked. As discussed in Sect. 4.3.1, if the solver status is "ok" and the termination condition is "optimal", then the global optimal solution is found. Otherwise, the error or output message is printed. In addition, the "time" library is used to attain the execution time of the model. The library is simply imported using `import time`.

Box 4.34

```
solver_name = Params.loc['solver_name','Val']

start_time = time.time()
results = pyo.SolverFactory(solver_name).solve(model)
end_time = time.time()

solver_status = results.solver.status
termination_condition = results.solver.termination_condition
run_time = end_time - start_time
```

```
7 if (solver_status = = SolverStatus.ok) and (termination_condition = =
                              TerminationCondition.optimal):
8    print(f"Solver: {Params.loc['solver_name','Val']}")
9    print(f"Solver Status: OK")
10   print("Termination Condition: Optimal")
11   print(f"run time (s): {run_time}")
12   pyo.assert_optimal_termination(results)
13 else:
14   print(f"Solver: {Params.loc['solver_name','Val']}")
15   print("Global optimal solution was not found.")
16   print(f"Solver Status: {solver_status}")
17   print(f"Termination Condition: {termination_condition}")
```

The solver name can be specified by the user in the "Params" sheet (line 1). Ensure that the specified solver is installed on the local machine for Pyomo to utilize it. The model is then solved (line 3), and if the solver status is "ok" and the termination condition is "optimal", line 12 is executed to load the optimal solution into the model variables. accessing the optimal value of the DVs is discussed in the next section.

4.3.16 Accessing the Results

After solving the model, one can access and analyze the optimal values of the DVs following the code presented in Box 4.35.

Box 4.35

```
power    = {}
reserve  = {}
onoff    = {}
start    = {}
shutdown = {}
emission = {}

for n in N:
    for g in G:
        power[(n,g)]    = {t: model.p[g,t,n].value for t in T}
        reserve[(n,g)]  = {t: model.r[g,t,n].value for t in T}
        onoff[(n,g)]    = {t: round(model.u[g,t,n].value) for t in T}
        start[(n,g)]    = {t: round(model.v[g,t,n].value) for t in T}
        shutdown[(n,g)] = {t: round(model.w[g,t,n].value) for t in T}
        emission[(n,g)] = {t: model.emsn[g,t,n].value +
                            round(model.v[g,t,n].value)*
```

	n1		n2		n3	
	g1	g4	g1	g3	g1	g2
1	20.000000	15.000000	20.000000	20.000000	20.000000	10.000000
2	37.007472	30.000000	36.912754	20.000000	34.261331	10.000000
3	22.007472	29.712238	36.614394	0.000000	22.263809	0.000000
4	20.000000	34.305853	32.990286	0.000000	33.786320	0.000000
5	0.000000	49.305853	34.226352	0.000000	39.547576	0.000000

Fig. 4.8 Optimal power generation values (first 5 rows)

```
                    GenData.loc[g,'StartUpEmission'] +
                    round(model.w[g,t,n].value)*
                 GenData.loc[g,'ShutDownEmission'] for t in T}
```

Lines 1–6 define a sample of six dictionaries that will be filled with the optimal values of their respective DVs. To find the optimal values, the `value` method must be used. For instance, to find the optimal power output of generator type 1 at time 3 in node 2 one must use `model.p['g1',3, 'n2'].value`. Once the dictionaries are created, they can be converted into a Pandas DataFrame for enhanced interpretability using `pd.DataFrame(power)`. The result is shown in Fig. 4.8.

These values can also be saved in an excel file and be used to plot various different charts to analyze the optimal solution.

4.4 Concluding Remarks

The aim of this section is to provide a step-by-step implementation of the constraints and equations presented in Chap. 3. To achieve such a goal, first the required libraries and input data were introduced. An Excel spreadsheet file is used to gather and organize all the required input files which is available in GitHub along with a Jupyter Notebook. Then, the linearization step is described in detail and implemented in Python. At the next stage, the OF is implemented and its different sections were explained. Afterwards, all the constraints were implemented and discussed. Finally, the model is solved using open-source solvers such CBC, HiGHS, or GLPK and the way to access the optimal values was mentioned.

References

1. Zhang R, Zhou J, Mo L, Ouyang S, Liao X (2013) Economic environmental dispatch using an enhanced multi-objective cultural algorithm. Electr Power Syst Res 99:18–29. https://doi.org/10.1016/j.epsr.2013.01.010
2. Soroudi A (2017) Unit commitment BT: power system optimization modeling in GAMS. In: Soroudi A (ed) Springer, Cham, pp 119–140
3. Xu T, Birchfield A, Gegner K, Shetye K, Overbye T (2017) Application of large-scale synthetic power system models for energy economic studies. https://doi.org/10.24251/HICSS.2017.386
4. Suess DT, Suess EA, Gregory SR (2009) Development of startup and shutdown permit limits based upon historical data from combustion sources monitored by continuous emission monitoring systems. Proc Air Waste Manag Assoc Annu Conf Exhib AWMA 5, 3232–3248. https://0201.nccdn.net/1_2/000/000/0a0/ce1/Development-of-Startup-and-Shutdown-Emission-Limits-Based-on-CEMS-Data.pdf
5. Maleki A, Askarzadeh A (2014) Optimal sizing of a PV/wind/diesel system with battery storage for electrification to an off-grid remote region: a case study of Rafsanjan, Iran. Sustain Energy Technol Assess 7:147–153. https://doi.org/10.1016/j.seta.2014.04.005
6. Cole W, Frazier AW, Cole W, Frazier AW (2020) Cost projections for utility-scale battery storage: 2020 update. Nat Renew Energy Lab
7. Carnegie R, Gotham D, Nderitu D, Preckel PV (2013) Utility scale energy storage systems benefits, applications, and technologies. https://www.purdue.edu/discoverypark/energy/assets/pdfs/SUFG/publications/SUFGEnergyStorageReport.pdf
8. Apribowo CHB, Sarjiya S, Hadi SP, Wijaya FD (2022) Optimal planning of battery energy storage systems by considering battery degradation due to ambient temperature: a review. Challeng New Perspect Batter 8:12. https://doi.org/10.3390/batteries8120290
9. Pena-Bello A, Barbour E, Gonzalez MC, Patel MK, Parra D (2019) Optimized PV-coupled battery systems for combining applications: impact of battery technology and geography. Renew Sustain Energy Rev 112:978–990. https://doi.org/10.1016/j.rser.2019.06.003
10. Energy Information Administration (2022) Annual energy outlook 2022. EIA. https://www.eia.gov/outlooks/aeo/data/browser/#/?id=8-AEO2022&region=0-0&cases=ref2022&start=2020&end=2050&f=A&linechart=ref2022-d011222a.75-8-AEO2022&ctype=linechart&sourcekey=0
11. Trading Economics (2023) EU carbon permits. Trading Economics. https://tradingeconomics.com/commodity/carbon. Accessed 26 Feb 2023
12. The World Bank (2023) Carbon pricing dashboard. The World Bank Group. https://carbonpricingdashboard.worldbank.org/map_data
13. Pyomo (2015) Accessing solver status and termination conditions. Pyomo. https://www.pyomo.org/blog/2015/1/8/accessing-solver. Accessed 20 Jan 2024

Chapter 5
Analyzing Optimization Results

Abstract Each operational constraint affects the optimal solution in a different way. The aim of this chapter is to first analyze the impact of each individual constraint on the optimal solution and then investigate their combined effect. To do so, the base model without the consideration of DR, renewable integration, EES, NZE, and reliability is solved and analyzed first. Then, each of these constraints is added to the base model, and the results are compared with the base model. Finally, the integrated case with all the operational constraints is solved.

5.1 Introduction

The previous chapter discussed the mathematical and Pyomo formulation of the economic and environmental optimization problem. Since the described model is modular, the modeler can add or remove each of the code blocks to understand the impact of each of the constraints on the optimization results. For instance, it is possible to have a plain model without considering reliability, demand response, renewables, and energy self-sufficiency. This chapter discusses the results of seven scenarios as shown in Table 5.1.

The following subsections presents the results of each of the scenarios presented in Table 5.1.

5.2 Base Case

The results of the baseline scenario (without reliability, demand response, storage, renewables, and net-zero assumptions) are brought in this section. To this end, Fig. 5.1 shows the performance of the generators in the baseline scenario.

As can be seen from Fig. 5.1, the model is capable of satisfying the electricity demand during the day except for the beginning and within the second peak hours (from 19 to 22). The hatched area shows the load loss, and the total amount of load

A. Ghadertootoonchi et al., *Energy System Modeling and Optimization*, SpringerBriefs in Energy, https://doi.org/10.1007/978-3-031-65906-5_5

Table 5.1 Seven scenarios discussed in Chap. 4

Scenario number	Description
1	Base case (without DR, renewables, EES, NZE, reliability)
2	Base case + DR
3	Base case + renewables
4	Base case + EES
5	Base case + NZE
6	Base case + reliability
7	Integrated case (base case with DR, renewables, EES, NZE, reliability)

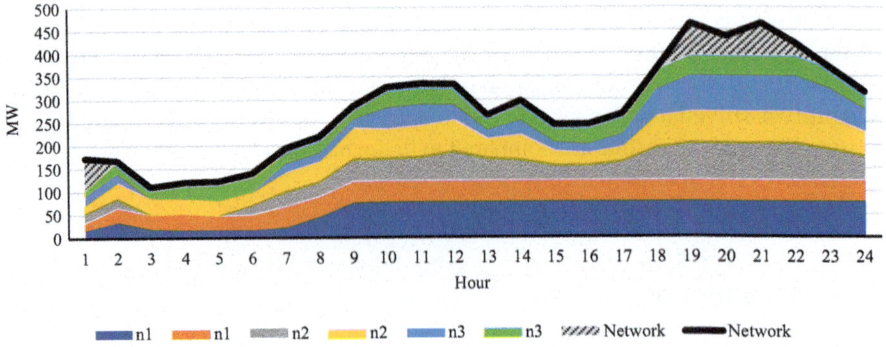

Fig. 5.1 Generators' performance in the base case scenario

lost is around 275 MWh which is less than the upper bounds introduced in Tables 4.1 and 4.4. In addition, no excess electricity is generated. Also, Fig. 5.1 shows that node 1 generators are providing the base load which is a result of the fact that the electricity demand in node 1 is the highest. Furthermore, the hourly emission is shown in Fig. 5.2.

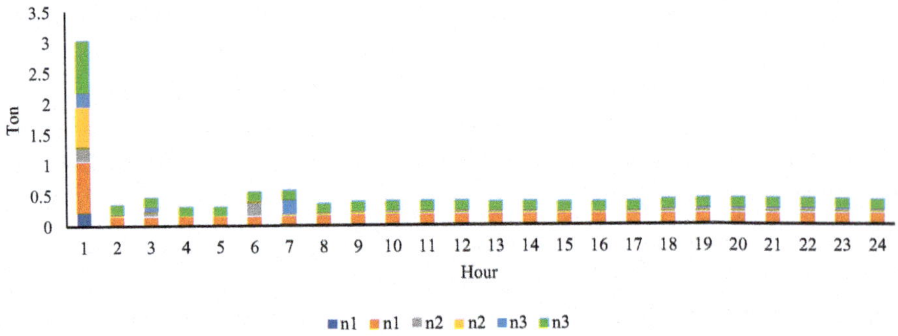

Fig. 5.2 Generators' hourly emission in the base case scenario

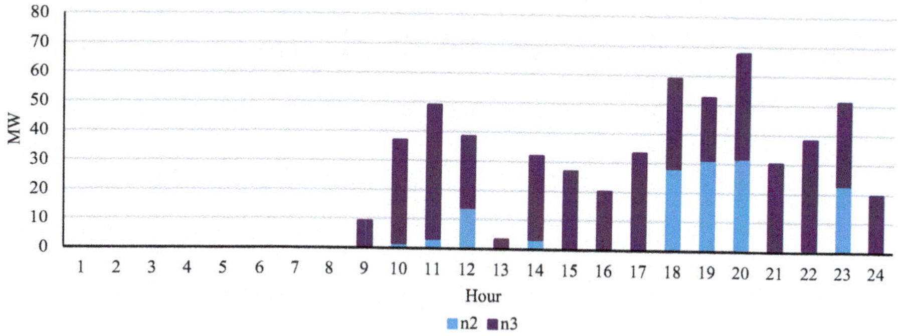

Fig. 5.3 Electricity flow between the nodes

Figure 5.2 shows that the highest emission occurred at the first timestep which is due to the startup of all the generators at $t = 1$. Then, the emissions fell until $t = 6$, at which generator 1 in node two is started again. After that, the emission experienced a plateau at about 0.42 Ton/h. Moreover, the electricity transmission between the nodes is shown in Fig. 5.3.

Figure 5.3 shows that the electricity flow is from nodes 2 and 3 to node 1. This is because node 1 has the highest electricity demand. In addition, as can be seen from Fig. 5.1, the total electricity generation of node 1 is at most equal to 125 MW which is 35 MW less than their 160 MW capacity. The reason for generating less electricity than the capacity is the quadratic cost function. The quadratic cost function makes the generation growth more expensive at high levels; Therefore, the model avoids pushing the generators to their maximum capacity. The objective values for the base case scenario are brought in Table 5.2.

Table 5.2 shows that the total cost of the network is positive because the costs exceed the income. In addition, the reserve income is zero, since the reliability (i.e., N − 1 contingency) and the reserve market are not considered in the base case. The sensitivity analysis of economic and emission costs of this case, considering different values for ω_c and ω_e, is presented in Fig. 5.4. The graph is developed considering (a) the emission price of 10 \$/Ton and (b) 100 \$/Ton.

As is clear from Fig. 5.4, when the emission price is 10 \$/Ton, the model is not sensitive to the cost weight unless it is 0. It is because the environmental costs are significantly lower than the economic costs. When the emission price is changed to 100 \$/Ton, the model's sensitivity to the weights has increased and the solutions for $\omega_c = 0, 0.1, 0.2$ are different. However, still when the cost weight is more than 0.2, the model is indifferent to the weights for the reasons mentioned previously. Additionally, it should be mentioned that when the cost weight is 0 and only the environmental costs are considered, the model increases the load loss to its upper limit (500 MWh), and when the cost weight is 1, the load loss is 275 MW. In the next section, the demand response effect is analyzed.

Table 5.2 Objective values for the baseline case

Parameter	Unit	Value
Objective value (considering weight factors)	–	12,282.36
Start-up and shutdown cost	$	830
Operation cost	$	114,404
Excess electricity cost (penalty)	$	0
Load loss cost (penalty)	$	54,987.83
Transmission cost	$	8048.29
Reserve income	$	0
Power income	$	153,834
Total cost	$	24,436.1
Weighted total cost	–	12,218.05
Start-up and shut-down emission cost	$	32.77
Operation emission cost	$	95.85
Total emission cost	$	128.62
Weighted emission cost	–	64.31

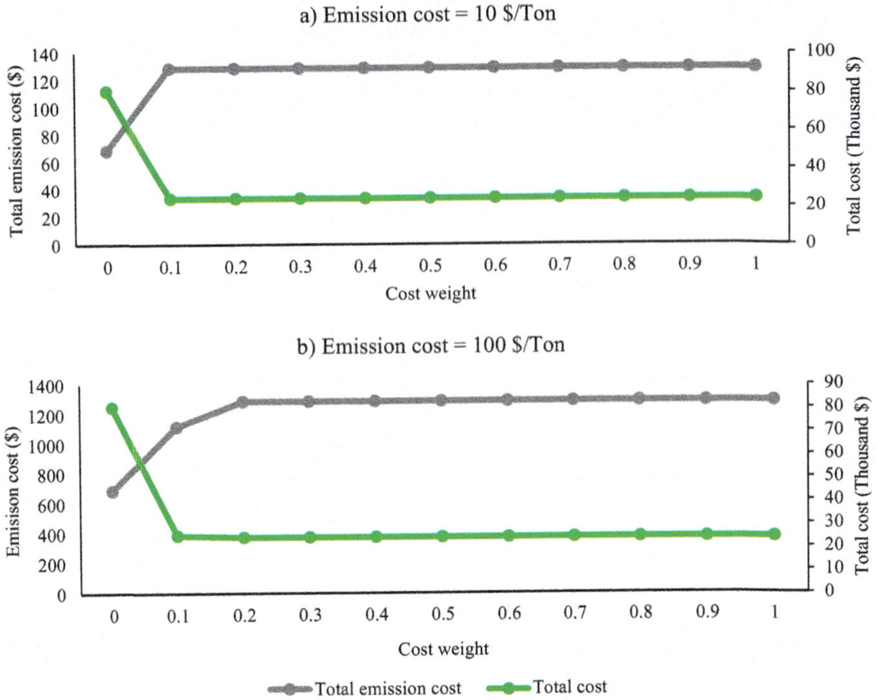

Fig. 5.4 The sensitivity of the economic and emission cost to the weight factors **a** emission cost is 10 $/Ton, **b** emission cost is 100 $/Ton

5.3 Base Case and DR

This section analyzes the scenario in which DR is integrated with the baseline scenario. At the beginning, it is considered that the DR is 25%, meaning that at each time step the model can increase or decrease the load by 25%. Then a sensitivity analysis is performed considering different values for the DR. The generators' performance in this scenario is shown in Fig. 5.5.

Figure 5.5 shows the optimization model has changed the load curve (the black dotted line) into the demand curve (the black line). A noticeable change is during the second peak (from 18 to 23). DR has reduced the load loss from 275 to 25 MWs, a 90.9% reduction. For further investigations, the economic and environmental cost curve considering different values for the demand side flexibility is brought in Fig. 5.6.

As is clear from Fig. 5.6, the more the demand-side flexibility, the less the load loss cost. Furthermore, it can be seen that the demand response environmental friendliness depends on the technologies available on the supply side. For instance, in the considered case, the growth in the demand side flexibility has led to a reduction in the load loss and therefore an increase in electricity generation. However, the impact of the generation growth on the emissions depends on the supply-side technologies. The reason for the sudden growth of the emissions when the demand response is 0.1 is that in this case the model has started up generator 3 twice (in other cases, generator 1 is started twice), and since the startup emission of the third generator is high, the emission is higher than the others. To reduce the emissions from electricity generation units, renewable sources can be used which are discussed in the next section.

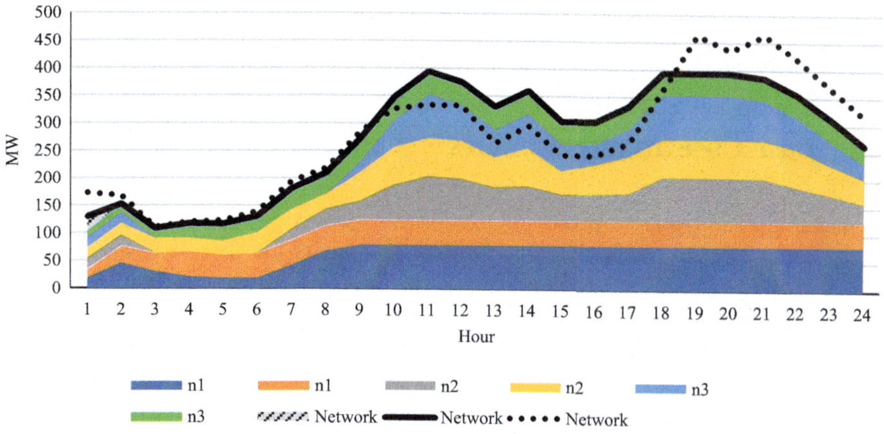

Fig. 5.5 Generators' performance considering a 25% demand response

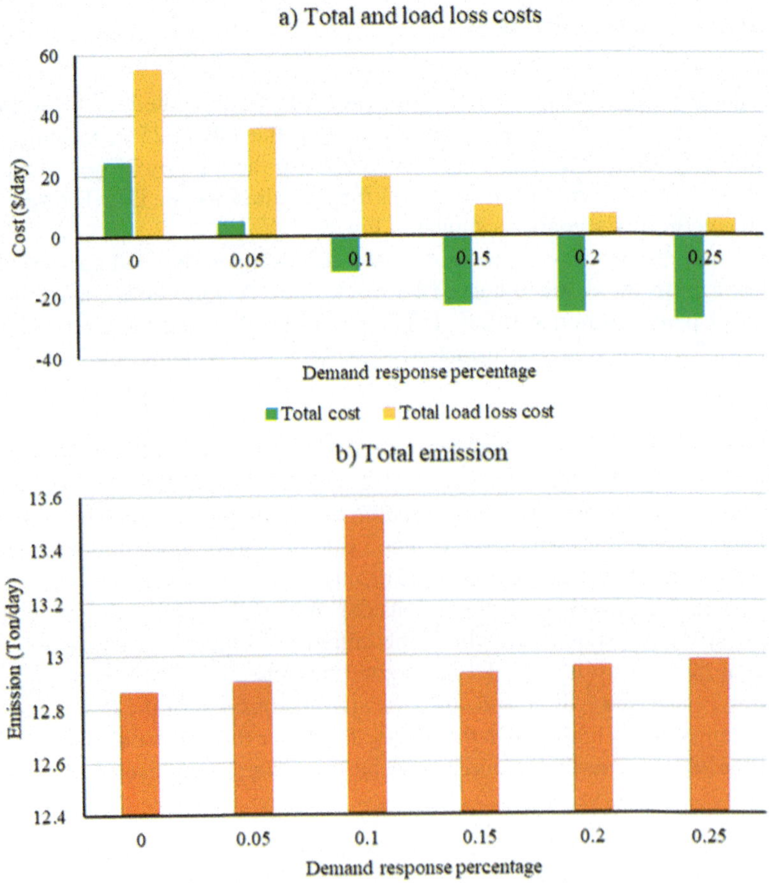

Fig. 5.6 Demand response effect on **a** the total and load loss costs and **b** the total emission cost

5.4 Base Case and Renewables

Renewable energies are a promising option for increasing the sustainability of the energy system and reducing its environmental impacts. Variable renewable sources like wind and solar have experienced significant growth during the last decade. In this section, the integration of solar energy into the base case scenario is investigated. The impact of solar energy integration on the performance of the thermal units is shown in Fig. 5.7.

Figure 5.7a, b show that the integration of renewable energies will reduce the utilization of thermal units. The reduction is 249 and 997 MWh/day in the 25 and 100 MW per node scenarios, respectively. In addition, the most noticeable drop belonged to the first generator (180 and 363 MWh/day, respectively) as it is the most expensive one. It is also worth mentioning that the integration of solar energy has

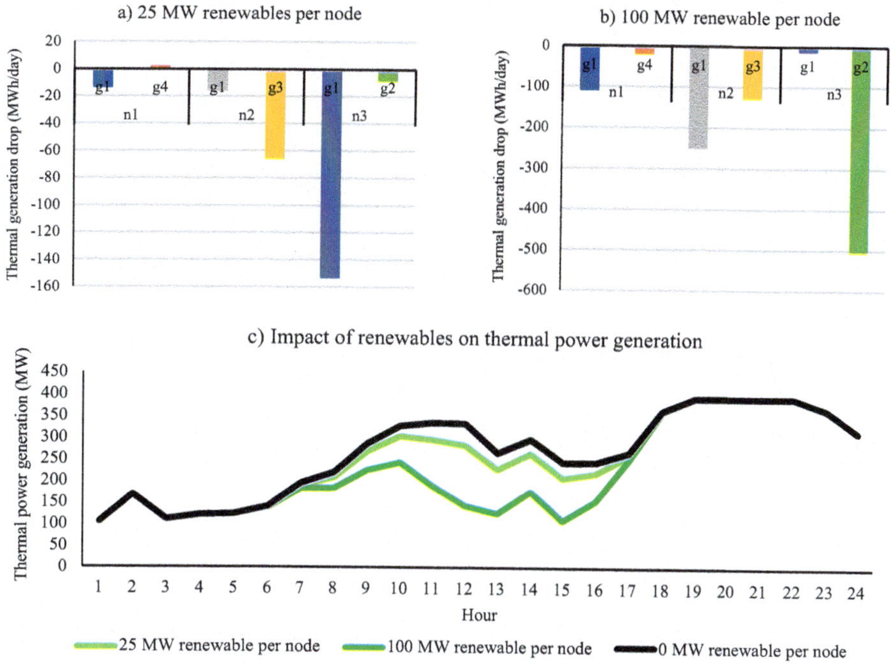

Fig. 5.7 Impact of solar energy integration on generators' performance **a** when the solar capacity is 25 MW per node and **b** 100 MW per node, and **c** the load curve alteration in presence of the solar energy

not changed the load loss as the load loss happens before or after the period within which renewable are available; Similarly, the amount of excess generation both for renewables and thermal units is zero.

With regard to the emissions, the results have revealed that the emission of the second case (25 MW per node) is higher than the baseline scenario which is due to the startup of the generators after the renewables phase-out. The emissions for the baseline, 25, and 100 MW per node are 12.9, 13.1, and 11.9 Ton/day respectively. This shows that the integration of solar energy into the electricity supply side does not necessarily alleviate the emissions and the reduction depends of the thermal generator's characteristics, especially their start up and shutdown emissions. In addition, variable energy sources are not dispatchable and their availability is highly uncertain. Therefore, storage systems play an important role when variable renewable sources are integrated into the energy system, the next section will analyze EES impact on the optimal operation of the network.

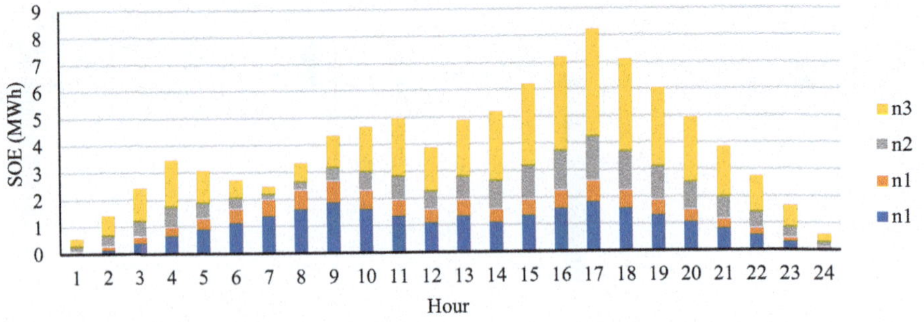

Fig. 5.8 The state of energy for storage units of different nodes

5.5 Base Case and EES

To analyze the impact of integrating EES into the base case, it is considered that four types of MW-scale batteries are available in the nodes; The characteristics of these units are brought in Table 4.3. Figure 5.8 shows the energy stored in every storage unit.

Based on Fig. 5.8, it is clear that the amount of energy stored in ESSs has a growing trend until the second peak at which the power price is highest, and after that, the stored electricity is discharged until the ESSs reach their minimum allowable level of stored energy. Figure 5.9a shows the change in the operation of thermal units in presence of ESSs and Fig. 5.9b represents their effect on the load loss and electricity flow in the network.

As depicted in Fig. 5.9, total generation has increased by 5.1 MWh/day and the load loss has decreased by 4.2 MWh/day. The reason that the load loss reduction is less than the generation growth is that 0.9 MWh/day is wasted due to the ESS and transmission lines' efficiency. In addition, the emissions have increased by 0.2 kg when ESS exists in the network while the total cost has witnessed a 4.6% reduction (from 24,436 to 23,313 $/day). The following section demonstrates the impact of energy self-sufficiency on the network performance.

5.6 Base Case and NZE

Nearly net-zero energy constraint will force the model to increase its electricity export to at least NZEC percentage of its total import during a day (see Eq. 3.53). This section will analyze the sensitivity of networks' optimal operation to different values of NZEC. To achieve this purpose, the NZEC is changed from 0 to 70% and the results are compared. For NZEC = 80 and 90% the model did not execute during the defined time limit of 1000 s, and for NZEC = 100% the model is infeasible meaning that all nodes cannot be energy self-sufficient at the same time during one day for

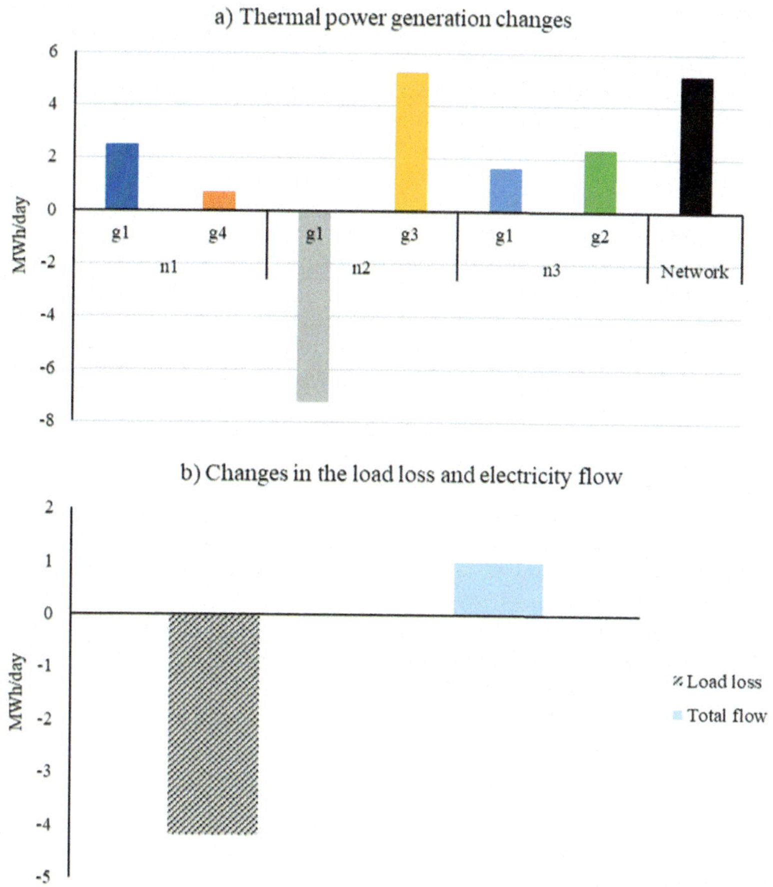

Fig. 5.9 The effect of ESSs on **a** the operation of the thermal units and **b** load loss and electricity flow

the considered case study. Figure 5.10 shows the amount of electricity exported and imported from the nodes.

Node 1 has the highest electricity demand and it has to import electricity, so it does not change the import and instead, increased the export to satisfy the nearly NZE constraint. In response, the other nodes have increased their import. Node 2 has the most expensive generators (1 and 3), so it prefers to not increase its export and instead increases its import from the other nodes so that the nearly NZE constraint is true for all nodes. Similarly, Node 3 has increased its import to absorb part of the growth in the export of Node 1. Figure 5.11 represents the export-to-import ratio (Exp/Imp) for all three nodes.

As Fig. 5.11 implies, node 1 imposes the binding nearly net zero energy constraint to the network, therefore its Exp/Imp ratio increases linearly with the respect to NZEC. Nodes 2 and 3 are not only NZ but also positive energy nodes as their

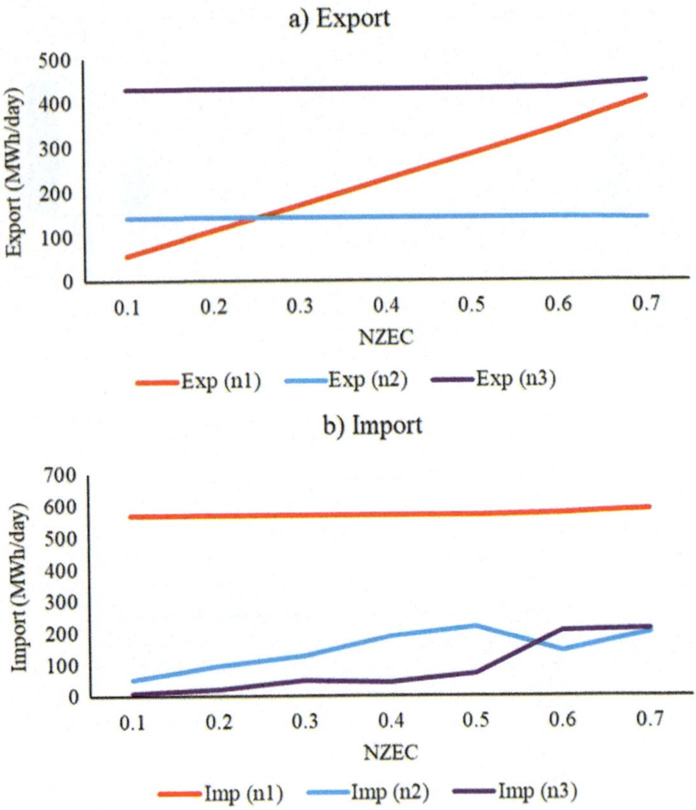

Fig. 5.10 **a** Electricity export and **b** import of different nodes

Exp/Imp ratios are greater than 1 (except for node 2 when $NZEC = 0.4, 0.5, 0.7$). Figure 5.12 depicts the cost and emission of the network considering different values for the NZEC.

Figure 5.12a shows that with the growth of NZEC, the total cost will increase; It is because of the surge in the system's cost while the income is approximately constant. For every 10% growth in NZEC, the network's cost is increased by approximately 4%. Similarly, the daily emission will grow with NZEC except for the case with $NZEC = 0.6$ in which the emission diminished; The drop is due to the reduced use of the second generator and the growth in using the third generator, which is more environmentally friendly, as shown in Fig. 5.13.

At the end of this section, the objective values for $NZEC = 0$ and $NZEC = 0.7$ are compared as shown in Fig. 5.14. As can be seen, the total cost of the network has increased by 29% and the highest contribution to the cost growth is related to the electricity transmission between nodes.

The next section will discuss the effect of reliability constraint on the network's performance.

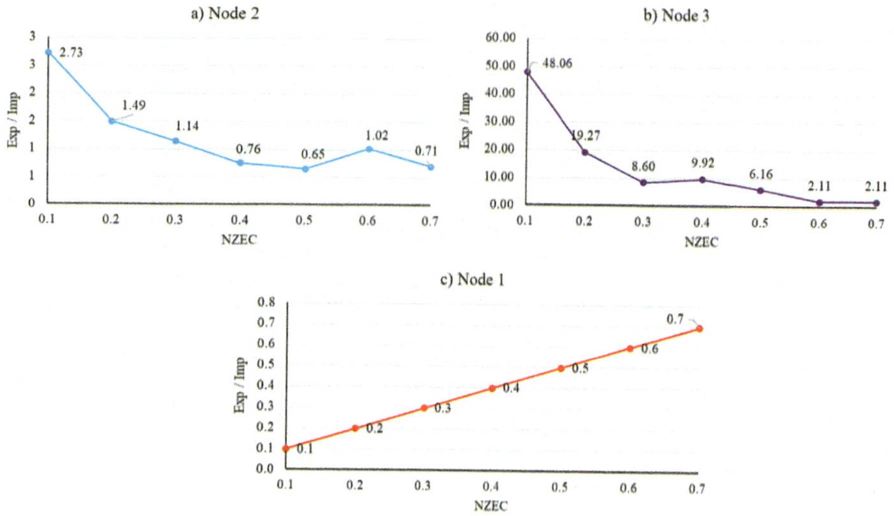

Fig. 5.11 Export to import ratio for **a** node 2, **b** node 3, and **c** node 1

5.7 Base Case and Reliability

N − 1 contingency is considered in this study to increase the reliability of the electricity network in the baseline scenario. N − 1 contingency states that the power generation of a generator must be less than the total reserve offered by other generators. It means that at every time step, the total offered reserve must exceed the maximum generation of all generators. Usually, some incentives are considered for the thermal units providing reserve in the form of a payment through the reserve market. Figure 5.15 shows (a) the total reserve offered by every generator during one day and (b) the total available hourly reserve in comparison to the maximum generation of all generators. It is assumed that $r_{max} = 0.75$ (see Eq. 3.69).

The reserve market integration will not change the load loss and excess power generation. Power generation is also similar to the case without the reliability; However, the profit has increased by 40,407.5 $ as a result of participating in the reserve market in which the price of reserve power is half of the power price.

Additionally, Fig. 5.15a show that the lowest contribution belongs to the second and fourth generators for which the ramp up rate is equal to 15 MW. On the other hand, the contribution of generators one and three are highest which is due to their larger ramp up rate (30 and 60 MW, respectively). Finally, Fig. 5.15b shows that the offered reserve is higher than the maximum power generation at each time step. The next section demonstrates the integrated case in which all the concepts discussed so far are considered.

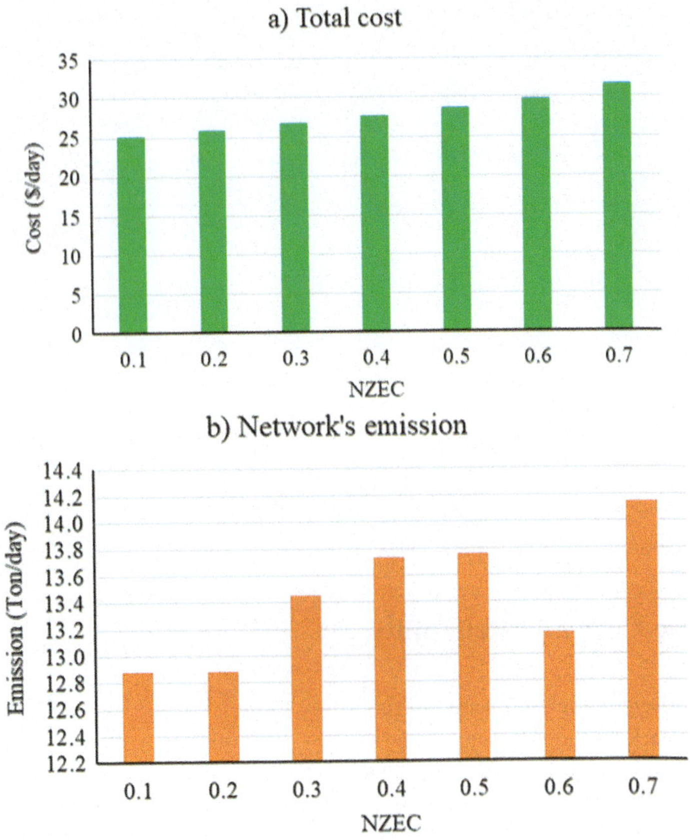

Fig. 5.12 The network's **a** total cost and **b** emission considering different values for NZEC

5.8 Integrated Case

In this section, the results of the integrated model are presented. In this scenario, the reliability ($r_{max} = 0.75$), NZE ($NZEC = 0.5$), DR ($\psi_{up} = \psi_{down} = 0.2$), Storage and renewables (25 MW per node) are considered. The share of each power source in covering the electricity demand is shown in Fig. 5.16.

Figure 5.16 shows that the node 1 generators provide the base load. Similar to the other scenarios, the integrated case has no excess electricity generation. Also, it is worth mentioning that the Exp/Imp ratios of nodes 1, 2, and 3 are 0.5, 0.5, and 29.96, respectively; It means that the third node is energy positive, while the others are partially zero energy nodes. The share of each energy source in covering the electricity demand is depicted in Fig. 5.17.

Figure 5.17 shows that 95% of the electricity demand is covered using the thermal generation units and generator type one located in node one is the highest utilized generator with a 25% share. Renewable (solar) energy covered 4% of the remaining

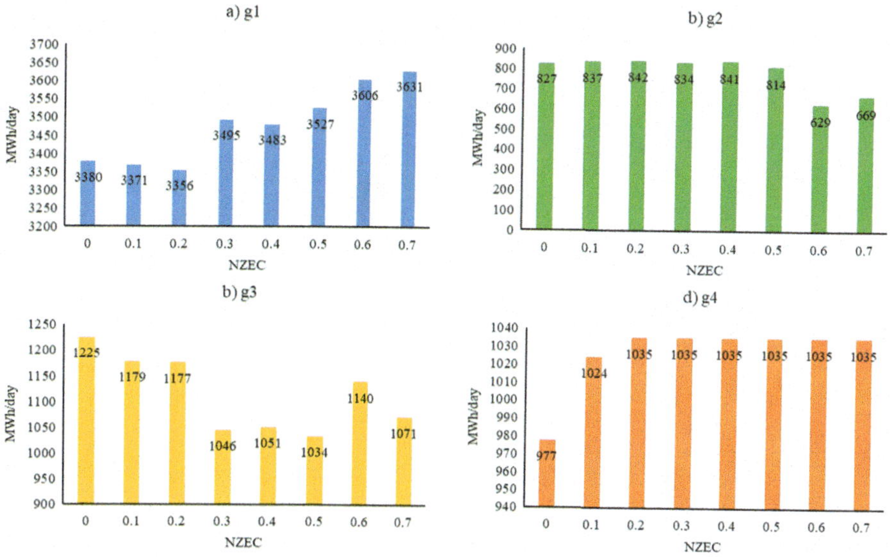

Fig. 5.13 The daily energy output of **a** generator type 1, **b** type 2, **c** type 3, and **d** type 4 considering different values of NZEC

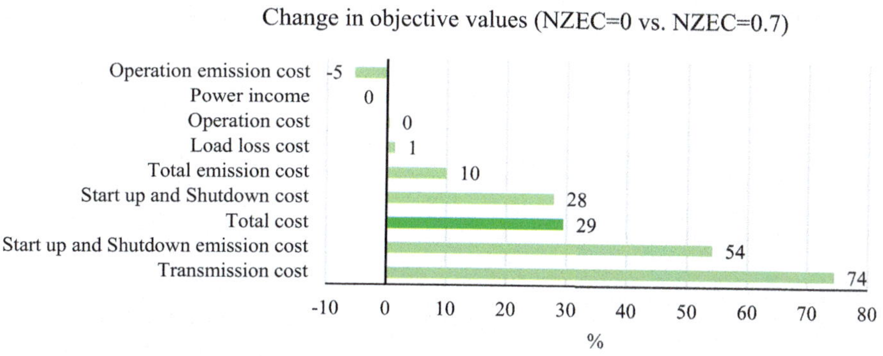

Fig. 5.14 Comparison of objective values for $NZEC = 0$ and $NZEC = 0.7$

5% whereas 1% of the demand is not covered or covered with EES system. The energy provided by energy storage systems is 11.45 MWh/day which is less than 1% of daily electricity demand. The objective values for the integrated and base case scenarios are compared in Table 5.3.

Table 5.3 clearly shows that the costs of the network diminished which mainly is a result of considering the reserve market and DR. In addition, the load loss penalty is reduced by around 88% which is due to the integration of renewables, storage systems, and DR.

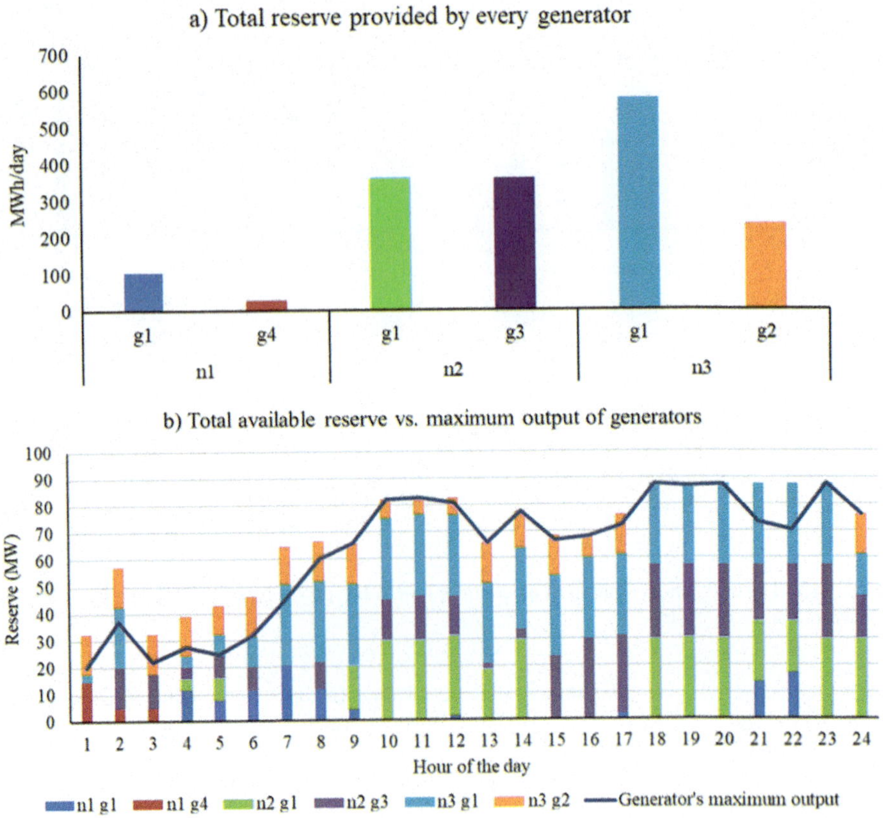

Fig. 5.15 **a** Total reserve offered by generators in a day, and **b** total available reserve in comparison with the maximum power generation

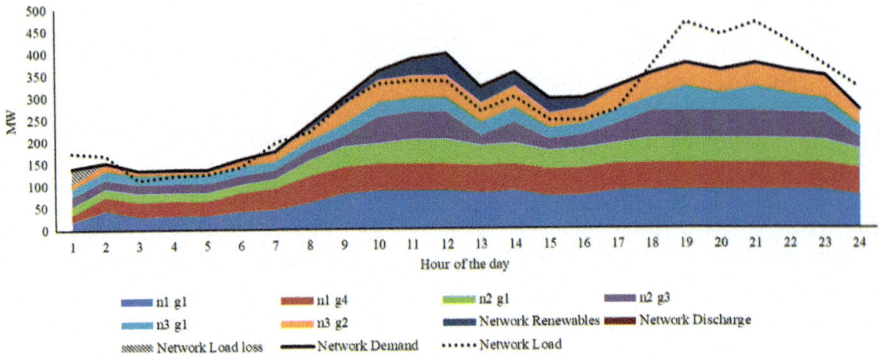

Fig. 5.16 Share of each power source in satisfying the electricity demand in the integrated scenario

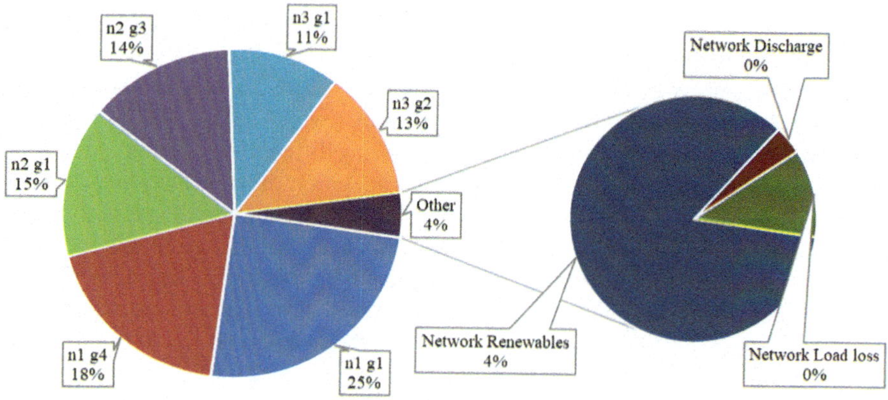

Fig. 5.17 Share of each energy source in covering the electricity demand of the network

Table 5.3 Objective values for the integrated scenario

Parameter	Unit	Value (integrated scenario)	Value (base case scenario)
Objective value (considering weight factors)	–	−27,211.9	12,282.36
Start-up and Shutdown cost	$	600	830
Operation cost	$	106,259.8	114,404
Excess electricity cost (penalty)	$	0	0
Load loss cost (penalty)	$	6756.3	54,987.83
Transmission cost	$	11,746.0	8048.29
Reserve income	$	19,984.2	0
Power income	$	159,912.5	153,834
Total cost	$	−54,534.6	24,436.1
Weighted total cost	–	−27,267.3	12,218.05
Start-up and shut-down emission cost	$	26.8	32.77
Operation emission cost	$	84.1	95.85
Total emission cost	$	110.8	128.62
Weighted emission cost	–	55.4	64.31

5.9 Concluding Remarks

This chapter aims to analyze the results of different scenarios using the economic and environmental optimization model developed in the previous chapter. To do so, a variety of different scenarios are considered and solved using open-source solvers such as HiGHS. Scenario definition using the developed Python code is straightforward since the code is modular and different constraints can be added or removed from the optimization model.

All the scenarios discussed in this section are easily accessible through the Microsoft Excel interface developed for the model which is available in GitHub. The reader is encouraged to run the code considering different parameters to become more familiar with their impact on the optimal operation of the electricity system.

There are several ways by which the presented ideas could be developed further. For instance, the heating demand of the nodes can also be considered and technologies such as boilers and CHP units can be integrated. Furthermore, multi-fuel thermal power and heating units can be considered. In addition, to increase the accuracy of the thermal units' cost function, the valve point effect can be added to the function. Moreover, the stochasticity associated with renewables, electricity demand, and electricity price can be taken into account using the idea of multi-stage optimization and robust optimization.

GPSR Compliance

The European Union's (EU) General Product Safety Regulation (GPSR) is a set of rules that requires consumer products to be safe and our obligations to ensure this.

If you have any concerns about our products, you can contact us on ProductSafety@springernature.com

In case Publisher is established outside the EU, the EU authorized representative is:

Springer Nature Customer Service Center GmbH
Europaplatz 3
69115 Heidelberg, Germany

The manufacturer's authorised representative in the EU is Springer Nature Customer Service Centre GmbH, Europaplatz 3, 69115 Heidelberg, Germany. If you have any concerns regarding our products, please contact ProductSafety@springernature.com

Printed and bound by CPI Group (UK) Ltd, Croydon, CR0 4YY
01/10/2026
02225985-0002